CONCEPTS IN ENTERPRISE

RESOURCE PLANNING

Second Edition

CONCEPTS IN ENTERPRISE

RESOURCE PLANNING

Second Edition

Ellen F. Monk
University of Delaware

Bret J. Wagner
Western Michigan University

THOMSON

COURSE TECHNOLOGY

Australia • Canada • Mexico • Singapore • Spain • United Kingdom • United States

Concepts in Enterprise Resource Planning, Second Edition

by Ellen F. Monk and Bret J. Wagner

Executive Editor
Mac Mendelsohn

Product Manager
Mirella Misiaszek

Development Editor
Amanda Brodkin

Production Editor
Philippa Lehar

Senior Acquisitions Editor
Maureen Martin

Senior Manufacturing Coordinator
Laura Burns

Senior Marketing Manager
Karen Seitz

Editorial Assistant
Jenny Smith

Cover Designer
Laura Rickenbach

Compositor
GEX Publishing Services

In memory of our colleague Majdi Najm. His support and friendship are sorely missed.

BRIEF CONTENTS

TABLE OF CONTENTS

This is a book about Enterprise Resource Planning (ERP) systems; it's also about how a business works and how information systems fit into business operations. More specifically, it's about looking at the processes that make up a business enterprise and seeing how ERP software can improve the performance of these business processes. ERP software is complicated and expensive. Unless a company uses it to become more efficient and effective in delivering goods and services to its customers, the ERP system will only be a drain on company resources.

When we wrote the first edition of this book in 2001 with our colleague, Joe Brady, it was because we couldn't find any books that were appropriate to use as a teaching text. At the time, there were no ERP textbooks—the only books available were professional books that covered ERP systems at a very high, strategic level. Our experience had shown us that:

1. Undergraduate business students don't always understand how businesses operate.

2. Advanced undergraduate students and even many MBA students do not truly grasp the problems inherent in unintegrated systems. These students also do not comprehend business processes and how different functional areas must work together to achieve company goals.

3. Students do not understand how an information system should help business managers make decisions.

Consequently, we set out to write a book that does the following:

1. Describes basic business functional areas and explains how they are related.

2. Illustrates how unintegrated information systems fail to support business functions and business processes that cut across functional area boundaries.

3. Demonstrates how integrated information systems can help a company prosper by improving business processes and by providing business managers with accurate, consistent, and current data.

The Approach of This Book

A key feature of our book is the use of the fictitious Fitter Snacker Company, a manufacturer of nutritious snack bars, as an illustrative example throughout the book. We show how Fitter Snacker's somewhat primitive and unintegrated information systems cause operational problems. We intentionally made the systems' problems easy to understand, so the student could readily comprehend them.

The first edition of *Concepts in Enterprise Resource Planning* was well received and was fairly successful in achieving its intended goal of providing a solid introduction to ERP systems. Since the first edition was released, we have been fortunate to continue our journey in learning more about ERP systems through SAP's University Alliance program. This program has provided the authors with access to SAP's R/3 software. It has also allowed us to interact with people at SAP and with SAP customers. These relationships help us better understand how the software supports business processes.

Because of our access to SAP software through the University Alliance, we chose to use SAP R/3 throughout the book as the sample system for ERP software. While many companies and educational institutes use software other than SAP for their ERP systems, the SAP examples in the book are provided as illustrations of how ERP concepts are handled by software. The important concept is that ERP systems should support business processes, regardless of what software product is used.

The second edition of *Concepts in Enterprise Resource Planning* adds a chapter on Human Resources Management. We also added a chapter on Process Modeling, Process Improvement, and ERP Implementation. We believe that faculty teaching ERP should avoid placing too much emphasis on large ERP implementation projects. In the post-Y2K environment, the important issue for most companies is determining how to leverage their investment in ERP software. This means that companies will be focusing on improving business processes, not managing large software implementation efforts. The examples of the SAP software included in the second edition use the current version of the system, R/3 Enterprise.

The second edition now has eight chapters:

- Chapter 1 explains the purposes for, and information systems requirements of, **main business functional areas**—Marketing and Sales, Supply Chain Management, Accounting and Finance, and Human Resources. This chapter also describes how a business process cuts across the activities within business functional areas and why managers need to think about making business processes work.

- Chapter 2 provides a short history of business computing and the **developments that led to today's ERP systems**. Chapter 2 concludes with an overview of ERP issues and an introduction to the SAP R/3 software.

- Chapter 3 describes the **Marketing and Sales functional area**, and it highlights the problems that arise with unintegrated information systems. To make concepts easy to understand, the Fitter Snacker running example is introduced. After explaining FS's problems with its unintegrated systems, we show how ERP can avoid these problems. SAP R/3 screens are used to illustrate the concepts. Because using ERP can naturally lead a company into ever-broadening integration, a discussion of customer-relationship management (CRM) concludes the chapter.

- Chapter 4 describes how ERP systems support **Supply Chain Management**—the coordinated activities of all the organizations involved in converting raw materials into consumer products on the retail shelf. As in Chapter 3, the problems caused by Fitter Snacker's unintegrated information system are explored, followed by a discussion of how ERP software could help solve these problems.

- Chapter 5 describes **Accounting and ERP systems.** This chapter is significantly revised from the first edition. It more clearly distinguishes between financial accounting (FI) and managerial accounting (CO) issues. The Enron collapse and Sarbanes-Oxley act were significant events that occurred since the first edition of this text was published. Both Enron and Sarbanes-Oxley are reviewed in detail, and their impact on information systems is discussed. The management controls and audit capabilities built into SAP are explored in detail.

- Chapter 6 is a new chapter that covers **Human Resource Management**. While the Human Resource software is the least integrated component of all ERP systems, it includes numerous processes that are critical to a company's success.

- Chapter 7 is a new chapter on **Process Modeling, Process Improvement, and ERP Implementation**. The chapter first presents flowcharting basics using a minimum number of symbols, followed by the highly structured EPC process model. Implementation issues conclude the chapter. Again, we feel that process improvement, not large-scale implementation, should be the focus in an introductory ERP course.

- Chapter 8, which was Chapter 6 in the first edition, covers **emerging topics in ERP** with an emphasis on Internet applications for ERP. Topics such as electronic commerce and application service providers are updated, while new material includes explanations of SAP NetWeaver and the emerging technology of RFIDs.

How Can You Use This Book?

The second edition adds material to supplement the first edition, while keeping the text at an introductory level. The book can be used in a number of ways:

- The book, or selected chapters, could be used for a three-week ERP treatment in undergraduate Management Information Systems, Accounting Information Systems, or Operations Management courses.

- Similarly, the book or selected chapters could be used in MBA courses, such as foundation Information Systems or Operations Management courses. Although the concepts presented here are basic, the astute instructor can build on them with more sophisticated material to challenge the advanced MBA student. Many of the exercises in the book require research for their solution, and the MBA student could do these in some depth.

- The book could serve as an introductory text in a course devoted wholly to ERP. It would provide the student with a basis in how ERP systems help companies to integrate different business functions. The instructor might use Chapter 8 as the starting point for teaching the higher-level strategic implications of ERP and related topics. The instructor can pursue these and related topics using his or her own resources, such as case studies and c: articles.

- Because of the focus on fundamental business issues and business processes, the book can be used in a sophomore level Introduction to Business course.

Except for a computer literacy course, we assume no particular educational or business background. Chapters 1 and 2 lay out most of the needed business and computing groundwork, and the rest of the chapters build on that base.

Features of This Text

To bring ERP concepts to life (and down to earth!) this book uses sales, manufacturing, purchasing, human resources, and accounting examples for the Fitter Snacker company. Thus, the student can see problems, not just at an abstract level, but within the context of a company's operations. We think this approach makes business problems and the role ERP can play in solving them easier to understand.

The book's exercises have the student analyze aspects of Fitter Snacker's information systems in various ways. The exercises vary in their difficulty; some can be solved in a straightforward way, and others require some research. Not all exercises need to be assigned. This gives the instructor flexibility in choosing which concepts to emphasize and how to assess students' knowledge. Some exercises explore FS's problems, and some ask the student to go beyond what is taught in the book and to research a subject. A solution might require the student to generate a spreadsheet, perform calculations, document higher-level reasoning, present the results of research in writing, or participate in a debate.

The book includes an additional element designed to bring ERP concepts to life: Another Look features, which are short, detailed case studies that focus on problems faced by real-world companies. Some of these cases include interviews with information systems managers who share their experiences with ERP.

We have illustrated ERP by showing how SAP R/3 would handle the problems discussed in the book. Screen shots of key SAP R/3 tools are shown throughout to illustrate ERP concepts. Many of the book's exercises ask the student to think about how a problem would be addressed using ERP software.

Supplemental Materials

This book is accompanied by the following materials:

- **Electronic Instructor's Manual**—The Instructor's Manual assists in class preparation by providing suggestions and strategies for teaching the text, chapter outlines, technical notes, quick quizzes, discussion topics, and key terms.

- **Solutions Manual**—The Solution's Manual contains answers to end-of-chapter questions and exercises.

- **Sample syllabi and course outline**—The Sample syllabi and course outlines are provided as a foundation to begin planning and organizing your course.

- **ExamView Test Bank**—ExamView allows instructors to create and administer printed, computer (LAN-based), and Internet exams. The Test Bank includes hundreds of questions that correspond to the topics covered in this text, enabling students to generate detailed study guides that include page references for further review. The computer-based and Internet testing components allow students to take exams at their computers, and also save the instructor time by grading each exam automatically. The Test Bank is also available in Blackboard and WebCT versions posted online at www.course.com.

- **PowerPoint Presentations**—Microsoft PowerPoint slides for each chapter are included as a teaching aid for classroom presentation, to make available to students on the network for chapter review, or to be printed for classroom distribution. Instructors can add their own slides for additional topics they introduce to the class.

- **Figure Files**—Figure and table files from each chapter are provided for your use in the classroom.

- **Hands-on SAP exercises**—Exercises are available for member institutions through the SAP University Alliance. These exercises use a database that was built for the fictitious Fitter Snacker company.

ACKNOWLEDGMENTS

Our thanks go out to our development editor, Amanda Brodkin, who learned how to work with authors who occasionally use too much jargon and have trouble meeting deadlines, and provided the critical eye that we needed to make our writing into what we imagined it was. We are grateful for the support and guidance of the entire MIS team at Course Technology, particularly product manager Mirella Misiaszek and production editor Philippa Lehar. We would not have been able to continue on our journey to understand ERP systems without the continued support of SAP America through its University Alliance program. We appreciate the efforts of Dan Pantaleo, Amelia Maurizio, and especially Heather Czech. Many SAP instructors have provided support above and beyond the call of duty, including Hilmer Hintz, William Welch, Steve Smith, Derek Purnell, "Jimmy" Flanagan, and Ashish Pradhan. Research support was provided by Lisa Kubota, Jaime Zorbo, Emily Best, Steve Isenhour, and Dina Durer. We also thank our reviewers Sarah Bee, Seattle University; Lauren Eder, Rider University; Cindy Joy Marselis, Temple University; Ted Weston, Colorado State University; and William Wagner, Villanova University, for insightful comments that pointed to needed improvements. In addition, we thank our interviewees, Linda Somers, Ellen Lepine, John Wheeler, and Pat Ryan with his colleagues from DuPont, as well as Phil Straniero and Don Scott, for their time and frankness. And finally, we thank our students, whose honesty and desire to learn have inspired us.

CHAPTER 1

BUSINESS FUNCTIONS, PROCESSES, AND DATA REQUIREMENTS

LEARNING OBJECTIVES

After completing this chapter, you should be able to:

- Name a business's main functional areas of operation.
- Differentiate a business process from a business function.
- Identify the kinds of data that each main functional area produces.
- Identify the kinds of data that each main functional area needs.
- Define integrated information systems and state why they are important.

INTRODUCTION

Enterprise Resource Planning (ERP) programs are core software used by companies to coordinate information in every area of the business. ERP (pronounced "E-R-P") programs help to manage company-wide business processes, using a common database and shared management reporting tools. A **business process** is a collection of activities that takes one or more kinds of input and creates an output that is of value to the customer. ERP software supports the efficient operation of business processes by integrating activities throughout a business, including sales, marketing, manufacturing, logistics, accounting, and staffing. In later chapters, you will learn how successful businesspeople use ERP programs to improve how work is done within a company. This chapter provides a background for learning about ERP software.

FUNCTIONAL AREAS AND BUSINESS PROCESSES

To understand ERP, you must first understand how a business works. Let's begin by looking at a business's areas of operation. These areas, called **functional areas of operation**, are broad categories of business activities.

Functional Areas of Operation

Most companies have four main functional areas of operation: **Marketing and Sales (M/S)**, **Supply Chain Management (SCM)**, **Accounting and Finance (A/F)**, and **Human Resources (HR)**. Each area comprises a variety of narrower **business functions**, which are activities specific to that functional area of operation. The business functions of each area are shown in Figure 1-1.

Functional area	Marketing and Sales	Supply Chain Management	Accounting and Finance	Human Resources
Business functions	Marketing of a product	Purchasing goods and raw materials	Financial accounting	Recruiting and hiring
	Taking sales orders	Receiving goods and raw materials	Cost allocation and control	Training
	Customer support	Transportation and logistics	Planning and budgeting	Payroll
	Customer relationship management	Scheduling production runs	Cash-flow management	Benefits
	Sales forecasting	Manufacturing goods		Government compliance
	Advertising	Plant maintenance		

FIGURE 1-1   Examples of functional areas of operation and their business functions

Historically, businesses have had organizational structures that separated the functional areas, and business schools have been similarly organized, so each functional area has been taught as a separate course. In a company separating functional areas in this way, Marketing and Sales might be completely isolated from Supply Chain Management, even though M/S sells what SCM procures and produces. Thus, you might conclude that what happens in one functional area is not closely related to what happens in others. Similarly, you might assume that one area's **information system (IS)** has no impact on the IS's of other functional areas. An IS includes the computers, people, procedures, and software that store, organize, and deliver information. As you will learn in this chapter, however, functional areas are interdependent, each requiring data from the others. The better a company can integrate the activities of each functional area, the more successful it will be in today's highly competitive environment.

Business Processes

Recently, managers have begun to think in terms of business processes rather than business functions. Recall that a business process is a collection of activities that takes one or more kinds of input and creates an output that is of value to the customer. The customer of a business process can be the traditional external customer (the person who buys the finished product), or it may be an internal customer (such as a colleague in another department). For example, what is sold through M/S is linked to what is procured and produced by SCM. This concept is illustrated in Figure 1-2.

Input/Functional area responsible for input	Process	Output
Request to purchase computer/Marketing and Sales	Sales order	Order is generated
Financial help for purchase/Accounting and Finance	Arranging financing in-house	Customer finances through the computer company
Technical support/Marketing and Sales	24-hour help line available	Customer's technical query is resolved
Fulfillment of order/Supply Chain Management	Shipping and delivery	Customer receives computer

FIGURE 1-2   Sample business processes related to the sale of a personal computer

Thinking in terms of business processes helps managers to look at their organization from the customer's perspective. For example, suppose that a customer wants to purchase a new computer. She wants information about the company's products so she can select a computer and various peripherals. She wants to place her order quickly and easily, and perhaps arrange for financing through the company. She expects quick delivery of a correctly configured, working computer, and she wants 24-hour customer support for any problems. The customer is not concerned about how the computer was marketed, or how its components were purchased, or how it was built, or the route the delivery truck took to get to her house. The customer wants the satisfaction of having a working computer at a reasonable price.

Businesses must always consider the customer's viewpoint in any transaction. What is the difference between a business function and a business process from the customer's point of view? Suppose the customer's computer is damaged during shipment. Because only one functional area is involved in accepting the damaged item, receipt of the return is a *business function* and is handled by the customer relationship management function of Marketing and Sales. Because several functional areas are involved in repair and return of the computer, the handling of the repair is a *business process*. Thus, the customer is dealing with many of the company's functional areas in her process of buying and obtaining a computer.

A successful customer interaction is one in which the customer (either internal or external) is not required to interact with each business function involved in the process. Successful business managers view their business operations from the perspective of a satisfied customer.

For the computer company to provide customer satisfaction, it must make sure that its functional areas of operation are integrated. For example, computer technology changes rapidly, and the hardware the computer company sells changes frequently. Therefore, people performing the sales function must have up-to-date information about computer configurations so they can provide the customer with accurate information; otherwise, the customer might order a computer that the company's manufacturing plant no longer produces. People performing the manufacturing function need to get the details of the customer's computer configuration quickly and accurately from those performing the sales function, so the right computer can be manufactured and shipped on time to the customer. If the customer is financing the computer through the computer company, then people performing the sales order function must gather information about the customer and process it quickly, so financing can be approved in time to support shipping the computer.

Sharing data effectively and efficiently between and within functional areas leads to more efficient business processes. Information systems can be designed so that functional areas share data. These systems are called **integrated information systems**. Working through this textbook will help you understand the benefits of integrated information systems and the problems that can occur when information systems are not integrated. Figure 1-3 illustrates the process view of business operations.

FIGURE 1-3   A process view of business

Businesses take inputs in the form of material, people, and equipment and transform these inputs into goods and services for the customer. Managing these inputs and the business processes effectively requires accurate and up-to-date information. For example, the sales function takes a customer's order, and the product's manufacture is scheduled by the production function. Logistics employees schedule and carry out the product's delivery. If raw materials are needed to make the product, production would prompt purchasing to arrange for their purchase and delivery. In that case, logistics would receive the material, verify its condition to accounting so that the vendor could be paid, and deliver the goods to production. Throughout, accounting would keep appropriate transaction records.

ANOTHER LOOK

Process-Oriented Management

Companies have been organized and managed by functional area for much of the 20th century, and significant progress has been made in improving the efficiency and effectiveness of businesses by improving the performance of the different business functions. But optimizing a business function does not mean that a company's business processes are being managed efficiently. For example, a company's purchasing department might reduce the costs to purchase raw materials by buying in large quantities. In isolation, it appears that purchasing is being very efficient. But the money saved by buying in large quantities may be more than offset by the cost of capital required to make the large purchase, as well as the costs to store the material and the risk that the material may become damaged, lost, stolen, or outdated before it can be used. If the purchasing department continually buys raw materials in large quantities, the company might have to invest in additional warehouse space.

Using a process-oriented view, the purchasing of raw materials is one part or function in the in-bound logistics process, which is the process of providing raw materials to the production line with the required quantity when needed. Deciding the quantity and timing of raw material purchases is just one part of this process, and this decision cannot be considered in isolation.

According to John Wheeler, CIO of Nova Chemicals, the change from function-oriented management to process-oriented management is itself a process—and a complicated one at that. "Business processes are more than a set of tasks or a workflow. They involve people with particular skill sets, information and tools. To be successful, they require the correct organizational culture," says Wheeler. He would characterize the transition of Nova Chemicals to a process orientation as being in a medium stage of maturity.

Some companies have moved almost completely to a process organizational structure. Instead of having departments with titles like Finance, Accounting, or Human Resources, they are organized around processes such as Order to Cash or Procurement to Payment.

continued

While functional departments still exist at Nova, the company has a process organizational structure as well, which consists of three core processes: Demand Chain Management, Supply Chain Management, and Manufacturing Management. These three processes are responsible for taking raw materials from suppliers and transforming them into products that are delivered to the customer. In addition to the three core processes, Nova has identified three enabling business processes: Human Resources (which includes corporate communications and investor relations), Treasury, and Information Technology. Nova has created process boards, consisting of managers from various functional areas, to manage these business processes. The boards determine best practices, which are guidelines that define how the processes are to operate. The process boards also define process metrics, which Nova defines as "clear measurements to manage process success." The process metrics are now driving the financial rewards (including raises and bonuses) to motivate employees to work in support of the process goals.

Having both a functional organization structure and a process structure can create the perception by employees that they have more than one manager. But according to Wheeler, "When people look at their daily tasks and ask the question 'Who do I work for?,' they start to realize that they work for many 'masters.' With well documented and managed business processes, employees can more clearly understand their place in the organization."

FUNCTIONAL AREAS AND BUSINESS PROCESSES OF A VERY SMALL BUSINESS

As an introduction, we will look at how business processes involve more than one functional area. To do this, we will look at a very small business as an example—a fictitious lemonade stand that you own. We will examine the business processes of the lemonade stand and see why coordination of the functional areas helps achieve efficient and effective business processes. You will see the role that information plays in this coordination and how integration of the information system improves your business.

Even though one person can run a lemonade stand, the operation of the business requires a number of processes. Coordinating the activities within different functional areas requires accurate and timely information.

Marketing and Sales

The functions of Marketing and Sales include developing products, determining pricing, promoting products to customers, and taking customers' orders. M/S also helps to create a sales forecast to ensure the successful operation of the lemonade stand.

For the most part, this is a cash business and does not require formal record keeping, but you still need to keep track of your customers so that you can send flyers to repeat customers or send occasional thank you notes. Thus, your records must show not only the amount of sales, but also must identify repeat customers.

Product development can be done informally in such a simple business; you gather information about who buys which kind of lemonade and note what customers say about each product. You also analyze historical sales records to spot trends that are not obvious.

Deciding whether to sell a product also depends on how much it costs to produce the product. For example, some customers might be asking for a sugar-free lemonade. You could determine whether the new lemonade could be profitably produced and sold by analyzing data from SCM, including production information (such as mixing container size, time required to mix) and materials management data (cost of lemons and sweetener).

Even though you run a cash business, good repeat customers are allowed to charge purchases—up to a point. Thus, your records must show how much each customer owes and his or her available credit. It is very important that the data are available and accurate at the time of a customer's credit request. Accounting and Finance records must be accessed as a part of the selling process, so the accounting function has a critical role to play in the sales process.

Supply Chain Management

The functions within Supply Chain Management include making the lemonade (production) and buying raw materials (purchasing). Production is planned so that, as much as possible, lemonade is available when needed, avoiding excess production of lemonade that must be liquidated. This planning requires sales forecasts from the M/S functional area, the accuracy of which will be improved if forecasts are based on historical sales figures. Thus, forecasts from M/S play an important role in the production planning process.

Production plans are also used to develop requirements for raw materials (bottled spring water, fresh lemons, and raw sugar) and packaging (cups, straws, and napkins). You must generate raw material and packaging orders from these requirements. If the forecasts are accurate, you will not lose sales because of material shortages, nor will you have excessive inventory that might spoil.

SCM and M/S must choose a recipe for each lemonade product sold. The standard recipe is a key input for deciding how much to order of each raw material, which is a purchasing function. This recipe also implies a need to keep good manufacturing records, so managers within the SCM functional area (working with those in A/F) can later compare how much it actually costs to make a glass of lemonade, versus how much the recipe *should* have cost.

Accounting and Finance

Functions within Accounting and Finance include recording raw data about transactions, including sales, raw material purchases, payroll, and receipt of cash from customers. **Raw data** are simply numbers collected from those areas without any manipulation, calculation, or arrangement for presentation. Those data are then summarized in meaningful ways to determine the profitability of the lemonade stand and to support decision making.

Note that data from Accounting and Finance are used by Marketing and Sales as well as by Supply Chain Management. The sales records are an important component of the sales forecast, which is used in making staffing decisions and in production planning. **Sales forecasts** are analyses that attempt to predict the future sales of a product. The records from accounts receivable, which you use to determine whether to grant credit to a particular customer, are also used to monitor the overall credit-granting policy of the lemonade stand. You want to be sure that you have enough cash on hand to purchase raw materials, as well as to finance purchasing new equipment, such as a lemon juicer.

Human Resources

Even a simple business must have employees who support the M/S and SCM functional areas. Such support implies a need to recruit, train, evaluate, and compensate employees—these are the functions of Human Resources.

At the lemonade stand, the number and timing of employees to hire depends on the level of lemonade sales. HR uses sales forecasts developed by the individual departments to plan personnel needs. A part-time helper might be needed at forecasted peak hours or days. How much should a part-time helper be paid? That depends on prevailing job market conditions, and it is HR's job to monitor those conditions.

Would increased sales justify hiring a part-time worker at the prevailing wage? Or, should you think about acquiring more automated ways of making lemonade, so that a person working alone could run the stand? These questions must be studied with input from SCM and A/F.

The lemonade stand, while a simple business, has many of the processes needed in larger organizations. These processes involve activities in more than one functional area; in fact, it is impossible to discuss the processes in one functional area without discussing the links to other functional areas. These connections invariably require sharing data. Systems that are integrated using ERP software provide the data sharing that is necessary between functional areas.

FUNCTIONAL AREA INFORMATION SYSTEMS

The lemonade stand provides simple examples of business processes and the functional activities required to support them. Next we will describe potential inputs and outputs for each functional area (refer back to Figure 1-1 to review inputs and outputs related to the sale of a personal computer). Note the kinds of data needed by each area and how people use the data. Also note the information systems' relationships between each functional area and process.

Marketing and Sales

The Marketing and Sales area needs information from all other functional areas to do its job. Customers communicate their orders to M/S by telephone, e-mail, fax, the Web, and so on. In the case of Web-based systems, customer and order data should be stored automatically in the information system; otherwise, data must be stored manually, by a person typing data into a keyboard or point-of-sale systems, or using a bar code reader or other device. Sales orders must be passed to SCM for planning purposes and to A/F for billing. In addition, sales order data can be used to analyze sales trends for business decision making. For example, M/S management might use a report showing the trend of a product's sales to evaluate marketing efforts and to determine strategies for the sales force.

M/S also has a role in determining product prices, which requires an understanding of the market competition and the costs of manufacturing the product. Pricing might be determined based on a product's unit cost, plus some percentage markup. For example, if a product costs $5 per unit to make, and management wants a 40 percent markup, the selling price must be $7 per unit. Where does the per-unit cost come from? Determining the cost of manufacturing a product requires information from Accounting and Finance which, in turn, relies on Supply Chain Management data.

To summarize, inputs for M/S include:

- Customer data
- Order data
- Sales trend data
- Per-unit cost

Outputs for M/S include:

- Sales strategies
- Product pricing

FIGURE 1-4 The Marketing and Sales functional area exchanges data with customers and with the Human Resources, Accounting and Finance, and Supply Chain Management functional areas

Supply Chain Management

Supply Chain Management also needs information from the various functional areas. Manufacturing firms develop production plans of varying length and detail, such as long-range, medium-range, and short-range plans. Each deals with different functional areas of the business. Examples of this planning might include expanding manufacturing capacity, hiring new workers, paying extra overtime for existing workers, and taking sales forecasts to plan manufacturing runs.

Production plans are based on information about product sales (actual and projected) that comes from Marketing and Sales. The Purchasing function bases its orders of raw materials on production plans, expected shipments, delivery lead times, and existing inventory levels. With accurate data about required production levels, raw material and packaging can be ordered as needed, and inventory levels can be kept low, saving money. On the other hand, if data are inaccurate or not current, manufacturing may run out of raw material or packaging; such a shortfall is called a **stockout**. Shortages of this type can shut down production and cause the company to miss delivery dates. To avoid stockouts, management might carry extra raw material and packaging, known as **safety stock**, which could result in an over-investment in inventory. The accuracy of the forecast determines the amount of safety stock that is required to reduce the risk of a stockout to an acceptable level. The less accurate the forecast, the more safety stock that is required. Accurate forecasting and production planning can reduce the need for extra inventory and manufacturing capacity.

Supply Chain Management records can provide the data needed by Accounting and Finance to determine how much of each resource (materials, labor, supplies, and overhead) was spent to make completed products in inventory.

Supply Chain Management data can support the M/S function by providing information about what has been produced and shipped. For example, some computer manufacturers, such as Gateway, have automated systems that call customers to notify them that their computer order has been shipped. Shipping companies, such as UPS and FedEx, provide shipment-tracking information on the Internet. By entering a tracking number, the customer can see each step of the shipping process by noting where the package's bar code was scanned. Thus, accurate and timely production information can support the sales process and increase customer satisfaction.

For long-range planning, management might want to see monthly reports showing sales and production figures. The data for such reports must come from the production and inventory data.

To summarize, inputs for SCM include:

- Product sales data
- Production plans
- Inventory levels

Outputs for SCM include:

- Raw material orders
- Packaging orders
- Resource expenditure data
- Production and inventory reports

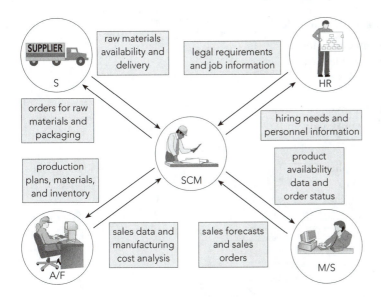

FIGURE 1-5 The Supply Chain Management functional area exchanges data with suppliers and with the Human Resources, Marketing and Sales, and Accounting and Finance functional areas

Accounting and Finance

Accounting and Finance needs information from all the other functional areas to complete its jobs accurately. A/F personnel record the company's transactions in the books of account. For example, they record accounts receivable when sales are made and cash receipts when customers send in payments. In addition, they record accounts payable when raw materials are purchased and cash outflows when they pay for materials. Finally, A/F personnel summarize the transaction data to prepare reports about the company's financial position and profitability.

People in other functional areas provide data to A/F: M/S provides sales data, SCM provides production and inventory data, and HR provides payroll and benefit expense data. The accuracy and timeliness of A/F data depend on the accuracy and timeliness of the data from the other functional areas.

M/S personnel require data from A/F to evaluate customer credit. If an order will cause the customer to exceed his or her credit limit, M/S should see that the customer's accounts receivable balance (the amount owed to the company) is too high and hold new orders until the customer's balance is lowered. If A/F is slow to record sales or customers' payments, then the accounts receivable balances are inaccurate. As a result, credit might be approved for customers who have already exceeded their credit limits and who might never pay off their accounts. Similarly, companies could deny credit to customers who actually owe less than their credit limit, potentially damaging their relationship with the customer.

To summarize, inputs for A/F include:

- Payments from customers
- Accounts receivables data
- Accounts payables data
- Sales data
- Production and inventory data
- Payroll and expense data

Outputs for A/F include:

- Payments to suppliers
- Financial reports
- Customer credit data

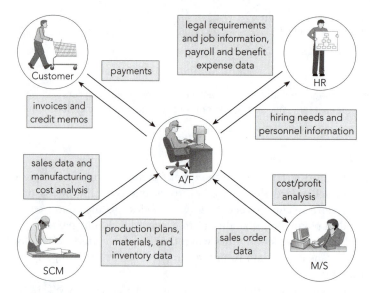

FIGURE 1-6 The Accounting and Finance functional area exchanges data with customers and with the Human Resources, Marketing and Sales, and Supply Chain Management functional areas

Human Resources

Like the other functional areas, HR also needs information from the other departments to do its job accurately. Tasks related to employee hiring, benefits, training, and government compliance are all the responsibilities of a human resource department. People in HR need accurate forecasts of personnel needs. HR also needs to know what skills are needed to perform a particular job and how much the company can afford to pay employees. These data come from all functional units.

State and federal laws require companies to observe many governmental regulations in recruiting, training, compensating, promoting, and terminating employees—and these regulations must be observed companywide. Usually, it is also HR's responsibility to ensure that employees receive training in a timely manner. HR also must ensure that employees get

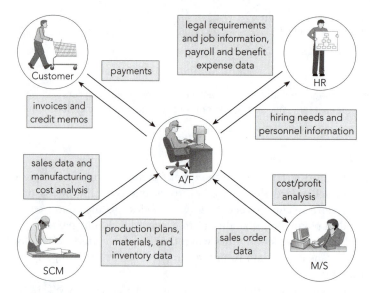

THE DEVELOPMENT OF ENTERPRISE RESOURCE PLANNING SYSTEMS

LEARNING OBJECTIVES

After completing this chapter, you should be able to:

- Identify the factors that led to the development of Enterprise Resource Planning (ERP) systems.
- Describe the distinguishing modular characteristics of ERP software.
- Discuss the pros and cons of implementing an ERP system.
- Summarize ongoing developments in ERP.

INTRODUCTION

In today's competitive business environment, companies try to provide customers with goods and services faster and cheaper than their competition. How do they do that? Often, the key is to have efficient, integrated information systems. Increasing the efficiency of information systems results in more efficient management of business processes. When companies have efficient business processes, they can be more competitive in the marketplace.

Until recently, most companies had unintegrated information systems that supported only the activities of individual business functional areas. Thus, a company would have a Marketing information

---

certified (and recertified) in key skills, such as materials handling and equipment operation. In addition, employees must be paid, and they must be considered for raises and bonuses. For these and other reasons, corporate HR needs timely and accurate data from other areas.

HR must create accurate and timely data for management use. For example, HR should maintain a database of skills required to do particular jobs and the prevailing pay rate. The data can be used to evaluate employees' performance and compensation. This analysis can help to prevent losing valued employees because of low pay.

To summarize, inputs for HR include:

- Personnel forecasts
- Skills data

Outputs for HR include:

- Regulation compliance
- Employee training and certification
- Skills database

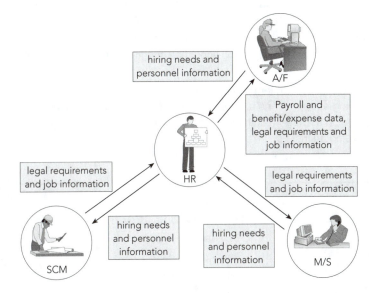

FIGURE 1-7   The Human Resources functional area exchanges data with the Accounting and Finance, Marketing and Sales, and Supply Chain Management functional areas

As shown in Figures 1-4 through 1-7, a significant amount of data is shared between the functional areas. The timeliness and accuracy of these data are critical to each area's success and the company's ability to make a profit and generate future growth. ERP software allows all the functional areas to share a common database so that accurate, real-time information is available. In the next chapter, we will trace the evolution of data management systems that led to ERP.

Chapter Summary

1. Companies that make and sell products have business processes that involve these basic functional areas: Marketing and Sales, Supply Chain Management, Accounting and Finance, and Human Resources. They perform these functions:

   - Marketing and Sales sets product prices, promotes products, takes customer orders, and creates sales forecasts.

   - Supply Chain Management develops production plans, orders raw materials from suppliers, receives the raw material into the facility, manufactures products, and ships products to customers.

   - Accounting and Finance records sales transactions, records customers' payments, records suppliers' invoices and payments to suppliers, and summarizes operational data in managerial reports.

   - Human Resources recruits, trains, compensates, and oversees the evaluation of employees.

2. A functional area is served by an information system. Information systems capture, process, and store data to provide information needed for decision making.

3. Employees working in one functional area need data from employees in other functional areas. Ideally, functional area information systems should be integrated, so shared data are accurate and timely.

4. Today business managers try to think in terms of business processes that integrate the functional areas, thus promoting efficiency and competitiveness. An important aspect of this is the need to share information between functions and functional areas. ERP software provides this capability by using a single common database.

Key Terms

Accounting and Finance (A/F)	Integrated information system
Business function	Marketing and Sales (M/S)
Business process	Raw data
Enterprise Resource Planning (ERP)	Safety stock
Functional areas of operation	Sales forecast
Human Resources (HR)	Stockout
Information system (IS)	Supply Chain Management (SCM)

Exercises

1. Distinguish between a business function and a business process. Provide an example to illustrate your points. Why do managers today try to think in terms of business processes rather than business functions?

2. Think of the last time you returned an item to a store. Describe how the process of your return cut across various functional lines in that business organization. Distinguish between the return process and the business function.

3. Assume you are the marketing manager of an up-and-coming firm that sells surfboards. What data do you need from Accounting and Finance and Supply Chain Management to do your job?

4. Assume you now are the chief financial officer of the company that sells surfboards. You need detailed records to report your financial status to the government. What functional areas provide that information? What information might you need to complete your report?

5. Assume you are the human resources manager of the company that sells surfboards. Your job is vital because your employees are the main asset to your firm, since each board is hand-crafted. What information do you need from each functional area to ensure proper staffing of the production line?

system, a Production information system, and so on, each with its own hardware, software, and methods of processing data and information.

Such unintegrated systems might work well within individual functional areas, but to achieve a company's goals, data must be shared among all the functional areas. When a company's information systems are not integrated, costly inefficiencies can result. For example, suppose two functional areas each have their own unintegrated information system. To share data, a clerk in one functional area needs to print out data from another functional area and then type the information into her area's information system. Not only does this data input take twice the time, it also significantly increases the chance for data entry errors. Alternatively, the process might be automated by having one information system write data to a file to be read by another information system. This would reduce the probability of errors, but it could only be done periodically (usually overnight or on a weekend) to minimize the disruption to normal business transactions. Because of the time lag in updating the system, the transferred data would rarely be up to date. In addition, data can be defined differently in different data systems, such as calling products by different part numbers in different systems. This can create further problems in timely and accurate information sharing between functional areas.

An Enterprise Resource Planning (ERP) system can integrate a company's operations by acting as a company-wide computing environment that includes a database that is shared by all functional areas. Such software can deliver consistent data across all business functions in real time. Real time refers to data and processes that are always current.

An ERP system can dramatically improve costs and operational efficiency. For example, IBM's Storage Systems division reports that it can now reprice its product inventory in five minutes rather than five days. Shipping a replacement part now takes three days instead of 22 days. Doing a customer credit check takes three seconds instead of 20 minutes.[1] These improvements can lead to lower costs and more satisfied customers. Toyota is saving $7 million per year by allowing its employees to access their own human resource records and giving group leaders access to shop floor information on employees through its new Human Resources ERP.[2]

This chapter will help you to understand how and why ERP systems came into being and what the future might hold for business information systems. The chapter follows this sequence:

- A review of the evolution of information systems and related causes for the recent development of ERP systems

- Discussion of the few ERP software vendors that dominate the market. The current industry leader, German software maker SAP AG, and its revolutionary software product, SAP R/3, are discussed as an example of an ERP system.

- Review of factors influencing a company's decision on whether to purchase an ERP system

- Description of ERP's benefits

- Overview of frequently asked questions related to ERP systems

- A discussion of the future of ERP software and its impact on the Internet, electronic commerce, and Web services

THE EVOLUTION OF INFORMATION SYSTEMS

It seems obvious today that a business should have integrated software to manage all functional areas. An integrated ERP system, however, is an incredibly complex hardware and software system that was not feasible until the 1990s. Current ERP systems evolved as a result of three things: the advancement of hardware and software technology (computing power, memory, and communications) needed to support the system; the development of a vision of integrated information systems; and the reengineering of companies to shift from a functional focus to a business process focus.

Computer Hardware and Software Development

Computer hardware and software developed rapidly in the 1960s and 1970s. The first practical business computers were the mainframe computers of the 1960s. While these computers began to change the way business was conducted, they were not powerful enough to provide integrated, real-time data for business decision making. Over time, computers got faster, smaller, and cheaper. The rapid development of computer hardware capabilities has been accurately described by Moore's Law. In 1965, Intel employee Gordon Moore observed that the number of transistors that could be built into a computer chip doubled every 18 months (see Figure 2-1). This meant that the capabilities of computer hardware were doubling every 18 months.

FIGURE 2-1   The actual increase in transistors on a chip approximates Moore's Law
Courtesy of Intel Corporation

During this time, computer software was also advancing to take advantage of the increasing capabilities of computer hardware. In the 1970s, relational database software was developed, providing businesses with the capability to store, retrieve and analyze large volumes of data. Spreadsheet software, a fundamental business tool today, became popular in the 1980s. With spreadsheets, managers could perform complex business analyses without having to rely on a computer programmer to develop custom programs.

The computer hardware and software developments of the 1960s and 1970s paved the way for the development of ERP systems.

Early Attempts to Share Resources

As PCs gained popularity in business in the 1980s, managers became aware that important business information was being stored on individual PCs, but there was no easy way to share the information electronically. Users needed a way to share expensive peripheral equipment (like printers and hard disks, which in the early 1980s were fairly expensive) and, more importantly, data.

By the mid-1980s, telecommunications developments allowed users to share data and peripherals on local networks. Usually, these networks were groups of computers connected to one another within a single physical location. This meant that workers could download data from a central computer to their desktop PC to do the actual work with the data.

This central computer-local computer arrangement is now called a **client-server architecture**. Servers (central computers) became more powerful and less expensive and provided scalability. **Scalability** means that when a piece of equipment's capacity is exceeded, its capacity can be increased by adding new hardware. In the case of a client-server network, the network can be made scalable by adding a server—thus extending the life of the hardware investment.

By the end of the 1980s, much of the hardware needed to support the development of ERP systems was in place: fast computers, networked access, and advanced database technology. Recall from Chapter 1 that ERP programs help to manage company-wide business processes using a common database. This common database holds a very large amount of data. The technology to hold that data in an organized fashion, and to retrieve that data easily, is the **database management system**, known as a **DBMS**. By the mid-1980s, the DBMS required to manage the development of complex ERP software existed. The final element required for the development of ERP software was the understanding and acceptance of the business community. Businesspeople did not yet have the vision to see the benefits of integrated information systems, nor did they possess the willingness to commit the resources to develop ERP software.

The Manufacturing Roots of ERP

The concept of an integrated information system took shape on the factory floor. Manufacturing software developed during the 1960s and 1970s, evolving from simple inventory tracking systems to **materials requirements planning (MRP)** software. MRP software allowed a plant manager to plan production and raw materials requirements by working backward from the **sales forecast**, the prediction of future sales. Thus, the manager first looked at Marketing and Sales' forecast of demand (what the customer wants); looked at the Production schedule needed to meet that demand; calculated the raw materials needed to meet production; and finally, projected raw materials purchase orders to suppliers. For a company with many products, raw materials, and shared production resources, this kind of projection was impossible without a computer to keep track of various inputs.

The basic functions of MRP could be handled by mainframe computers using sequential file processing. **Electronic data interchange (EDI)**, the direct computer-to-computer exchange of standard business documents, allowed companies to handle the purchasing process electronically, avoiding the cost and delays resulting from paper purchase order and invoice systems. The functional area now known as Supply Chain Management (SCM) began with the sharing of long-range production schedules between manufacturers and their suppliers.

Management's Impetus to Adopt ERP

The hard economic times of the late 1980s and early 1990s caused many companies to downsize and reorganize. These company overhauls were a stimulus to ERP development. Companies needed to find some way to avoid the following kind of situation, which they had tolerated for a long time.

Imagine you are the CEO of a clothing manufacturing company. Your company is profitable and is keeping pace with the competition, but your IS is unintegrated and inefficient (as are the systems of your competitors). You've learned to live with this kind of inefficiency: Your Marketing and Sales department creates a time-consuming paper trail for negotiating and making a sale. To schedule factory production, however, your Manufacturing manager needs accurate, timely information about actual and projected sales orders from the M/S manager. Without such information, the Manufacturing manager must guess which products to produce—and how many of them to produce. To keep goods moving through the production line, the manager often does guess. Sometimes the guess overestimates demand for some garments, and sometimes it underestimates demand.

Overproduction of a certain garment might mean your company is stuck with garments for which there is no market, or you might face a diminishing market due to style changes or seasonal demand. When you store the garments, waiting for a buyer, you incur warehouse expense. On the other hand, underproduction of a certain garment might result in garments not being ready for delivery when a salesperson promised, resulting in unhappy customers and cancelled orders. If you try to "catch up" on orders, you'll have to pay factory workers overtime pay, or resort to the extra expense of rapid-delivery shipments.

The management of large companies decided they could no longer afford the type of inefficiencies illustrated in the previous paragraphs, inefficiencies that occurred as the result of a functional model of business organization. This model, illustrated in Figure 2-2, had deep roots in U.S. business, starting with the General Motors organizational model developed by Alfred P. Sloan in the 1930s. The functional business model directed information to top management without sharing data between functional areas.

The functional model was very useful for decades, and was successful in the United States, where there was limited competition and where flexibility and rapid decision-making were not required to be successful. In the rapidly changing markets of the 1990s, however, the functional model led to top-heavy and overstaffed organizations incapable of reacting quickly to change. The time was right to view a business as a set of cross-functional processes, as illustrated in Figure 2-3.

In a process-oriented company, the flow of information and management activity are "horizontal" across functions, in line with the flow of materials and products. This horizontal flow promotes flexibility and rapid decision-making. Michael Hammer's 1993 landmark book, *Reengineering the Corporation: A Manifesto for Business Revolution,* stimulated managers to see the importance of managing business processes. Books like Hammer's, along with the difficult economic times of the late 1980s, led to a climate in which managers began to view ERP software as a solution to business problems.

In the next section, you will learn about the development of the first ERP software. SAP was the first company to develop software for ERP systems and is the current market leader in ERP software sales, with 54 percent of the market as of early 2004.[3]

FIGURE 2-2   Information and material flows in a functional business model

FIGURE 2-3   Information and material flows in a process business model

ERP SOFTWARE EMERGES: SAP AND R/3

In 1972 five former IBM systems analysts in Mannheim, Germany—Dietmar Hopp, Claus Wellenreuther, Hasso Plattner, Klaus Tschira and Hans-Werner Hector—formed *System-analyse und Programmentwicklung* (Systems Analysis and Program Development, or "S-A-P"). At this time, the computer industry was quite different from today. IBM controlled the computer market with its 360 Mainframe computer, which had only 512K of main memory. In this mainframe computer environment, the SAP founders recognized that all companies developing computer software faced the same basic business problems, and each developed unique, but similar, solutions for their needs in payroll processing, accounting, materials management, and other functional areas of business. SAP's goal was to develop a standard software product that could be configured to meet the needs of each company. SAP's concept from the beginning was to set standards in information technology, according to founder Dietmar Hopp.[4] In addition, the founders wanted data available in real time, and they wanted users to work on a computer screen, not with voluminous printed output. These goals were lofty and forward-looking for 1972, and it took almost 20 years to achieve them.

SAP Begins Developing Software Modules

Before leaving IBM, Plattner and Hopp had worked on an order processing system for the German chemical company ICI. The order processing system was so successful that ICI managers also wanted a materials and logistics management system—a system for handling the purchase, receiving, and storage of materials that could be integrated into the new order processing system. During their work at ICI, Plattner and Hopp had already developed the idea of modular software development. Software **modules** are individual programs that can be purchased, installed, and run separately, but that all extract data from the common database. Instead of giving Plattner and Hopp the new ICI project, IBM made them traveling software experts, removing them from hands-on development duties. Plattner and Hopp approached Claus Wellenreuther, an expert in financial accounting who had just left IBM, to form their own company.

When Plattner, Hopp, and Wellenreuther established SAP on April 1, 1972, they could not afford to purchase their own computer. Their first contract, with ICI to develop the follow-on materials and logistics management system, included access to ICI's mainframe computer at night to develop the software package, a practice they repeated with other clients until they acquired their first computer in 1980. At ICI, the SAP founders developed their first software package, variously called System R, System RF (for real-time financial accounting) and R/1.

To keep up with the ongoing development of mainframe computer technology, in 1978 SAP began developing a more integrated version of its software products, called the R/2 system. In 1982, after four years of development, SAP released its R/2 mainframe ERP software package.

Sales grew rapidly in the 1980s, and SAP extended its software's capabilities and expanded into international markets. This was no small task, because the software had to be able to accommodate different currencies, accounting practices, and tax laws.

By 1988, SAP had established subsidiaries in numerous foreign countries, established a joint venture with consulting company Arthur Andersen, and sold its 1,000th system. SAP also became SAP AG, a publicly traded company.

SAP R/3

In 1988, SAP realized the potential of client-server hardware architecture and began development of its **R/3** system to take advantage of client-server technology. The first version of SAP R/3 was released in 1992. Each subsequent release of the SAP R/3 software contained new features and capabilities. The client-server architecture used by SAP allowed R/3 to run on a variety of computer platforms, including UNIX and Windows NT. The SAP R/3 system was also designed using an **open architecture** approach. In open architecture, third-party software companies are encouraged to develop add-on software products that can be integrated with existing software. The open architecture also makes it easy for companies to integrate their hardware products, such as bar code scanners, personal digital assistants (PDAs), cell phones, and global information systems (GIS) with the SAP R/3 system.

New Directions in ERP

In the late 1990s, the Year 2000, or Y2K, problem motivated many companies to move to ERP systems. As it became clear that the date turnover from December 31, 1999 to January 1, 2000 would wreak havoc on information systems, companies searched for ways to consolidate data, and ERP systems provided one solution.

The Y2K problem originated from programming shortcuts made by programmers in the preceding decades. With memory and storage space a small fraction of what it is today, early programmers developed software that used as few computer resources as possible. To save memory, programmers in the 1970s and 1980s typically wrote programs that only used two digits to identify a year. For example, if an invoice was posted on October 29, 1975, the programmer could just store the date as 10/29/75 rather than 10/29/1975. While this may not seem like a big storage savings, with millions of transactions needing to be manipulated, it adds up. These programmers never imagined that software written in the 1970s would still be running major companies and financial institutions in 1999. These old systems were known as **legacy systems**. Many companies were faced with a choice: pay programmers millions of dollars to correct the Y2K problem in their old, limited software—or invest in an ERP system to not only solve the Y2K problem, but potentially gain better management of their business processes as well. Thus, the Y2K problem led to a dramatic increase in business for ERP vendors in the late 1990s. The rapid growth of the 1990s was followed by an ERP slump starting in 1999. By 1999, companies were either in the final stages of an ERP implementation or modification of their existing software. Many companies that had not yet decided to move to a Y2K-compliant ERP system waited until after the new millennium to upgrade their information systems.

By 2000, SAP AG had 22,000 employees in 50 countries and 10 million users at 30,000 installations around the world. By that time, SAP had competition in the ERP market, namely from Oracle and PeopleSoft (PeopleSoft expanded its offerings through the acquisition of ERP software vendor J. D. Edwards in 2003). In late 2004, Oracle succeeded in its bid to take over PeopleSoft. The implications of this merger were unknown at the time of this book's publication.

PeopleSoft

PeopleSoft was founded by David Duffield, a former IBM employee who, like SAP's founders, faced opposition from IBM for his ideas. PeopleSoft started with software for Human Resources and Payroll Accounting, and achieved considerable success, even with companies that already were using SAP for Accounting and Production. PeopleSoft's success caused SAP to make significant development to its Human Resources module. PeopleSoft expects to strengthen its offerings in the supply chain area with its acquisition of J.D. Edwards.

Oracle

Oracle is SAP's biggest competitor. Oracle began in 1977 as Software Development Laboratories (SDL). Its founders, Larry Ellison, Bob Miner, and Ed Oates, won a contract from the Central Intelligence Agency (CIA) to develop a system, called Oracle, to manage large volumes of data and extract information quickly. While the Oracle project was cancelled before a successful product was developed, the three founders of SDL saw the commercial potential of a relational database system. In 1979, SDL became Relational Software, Inc. and released its first commercial database product. The company changed its name again, to Oracle, and in 1986 released the client-server Oracle relational database. The company continued to improve its database product, and in 1988 released Oracle Financials, a set of financial applications. The financial applications suite of modules included Oracle Financials, Oracle Supply Chain Management, Oracle Manufacturing, Oracle Project Systems, Oracle Human Resources, and Oracle Market Management. Oracle Financials was the beginning of what would become Oracle's ERP product.

The concepts of an enterprise resource planning system are similar for large vendors, such as SAP, Oracle, and PeopleSoft, as they are for the many smaller vendors of ERP software. Because of SAP's leadership in the ERP industry, this textbook focuses primarily on the SAP software. Keep in mind that most other software vendors provide similar functionality, with some having strengths in certain areas.

SAP R/3 Enterprise

The latest versions of ERP systems by SAP and other companies allow all business areas to access the same database, as shown in Figure 2-4, eliminating redundant data and communications lags. Perhaps most importantly, the system allows data to be entered once, and then used throughout the organization. In information systems, errors most frequently occur where humans interact with the system. ERP systems ensure that data are entered only once, where they are most likely to be accurate. For example, with access to real-time stock data, a salesperson taking an order can confirm the availability of the desired material. When the salesperson enters the sales order into the system, the order data are immediately available to Production, so Manufacturing can update production plans and Materials Management can plan the delivery of the order. If the sales order data are entered correctly by the salesperson, then SCM personnel are working with the same, correct data. The same sales data are also available to Accounting for preparing invoices.

Figure 2-5 shows the major functional modules in the current SAP R/3 system (known as release 4.7 or Enterprise) and depicts how the modules provide integration. The basic functions of each of the modules are described after the figure.

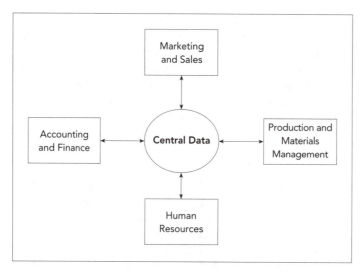

FIGURE 2-4 Data flow within an integrated information system

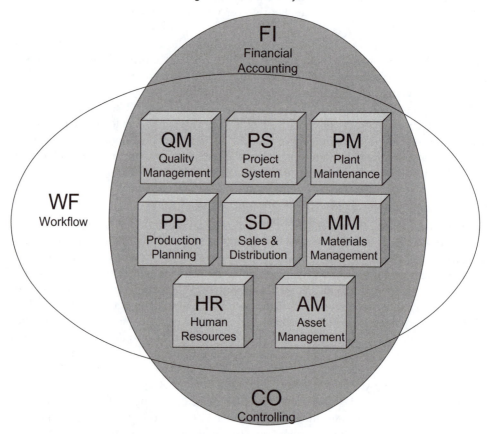

FIGURE 2-5 Modules within the SAP R/3 integrated information systems environment
Courtesy of SAP AG

- The **Sales and Distribution (SD)** module records sales orders and scheduled deliveries. Information about the customer (pricing, how and where to ship products, how the customer is to be billed, and so on) is maintained and accessed from this module.
- The **Materials Management (MM)** module manages the acquisition of raw materials from suppliers (purchasing) and the subsequent handling of raw materials inventory, from storage to work-in-progress goods to shipping of finished goods to the customer.
- The **Production Planning (PP)** module maintains production information. Here production is planned and scheduled, and actual production activities are recorded.
- The **Quality Management (QM)** module plans and records quality control activities, such as product inspections and material certifications.
- The **Plant Maintenance (PM)** module manages planning for preventative maintenance of plant machinery and managing maintenance resources, so equipment breakdowns are minimized.
- The **Asset Management (AM)** module helps the company to manage fixed-asset purchases (plant and machinery) and related depreciation.
- The **Human Resources (HR)** module facilitates employee recruiting, hiring, and training. This module also includes payroll and benefits.

Two financial modules, FI and CO, are shown as encompassing the modules described above. That is because nearly every activity in the company has an impact on the financial position of the company.

- The **Financial Accounting (FI)** module records transactions in the general ledger accounts. It is used to generate financial statements for external reporting purposes.
- The **Controlling (CO)** module is used for internal management purposes. Here, the company's manufacturing costs are assigned to products and to cost centers, so that the profitability of the company's activities can be analyzed. The CO module supports managerial decision making.
- The **Project System (PS)** module allows the planning for and control over new R&D, construction, and marketing projects. This module allows for costs to be collected against a project, and it is frequently used to manage the implementation of the SAP R/3 system. Companies use PS to manage build-to-order items. These build-to-order items are low volume, high complex products like ships and aircrafts. Kawasaki Industries uses PS to manage their building of subway cars.
- The **Workflow (WF)** module is not a module that automates a specific business function. Rather, it is a set of tools that can be used to automate any of the activities in SAP R/3. It can perform task-flow analysis and prompt employees (by e-mail) if they need to take action. Workflow is ideal for business processes that are not daily activities (like preparing customer invoices) but that occur frequently enough to be worth the effort to implement workflow.

To summarize: ERP integrates business functional areas with one another. Before ERP, each functional area operated independently, using its own information systems and ways of recording transactions. ERP software also makes management reporting and decision-making faster and more uniform throughout an organization. In addition, ERP promotes thinking about corporate goals, as opposed to thinking only about the goals of a single department or functional area. When queried on their reasons behind implementing ERP systems, the overriding answer given by top management is *control*. With the capability to see integrated data on their entire company's operation, managers use ERP systems for the control it provides, allowing them to set those corporate goals correctly.

SAP R/3 Software Implementation

A truly integrated information system entails integrating all functional areas, but for various reasons, not all companies that use SAP use all of the SAP R/3 modules. For example, a company without factories wouldn't choose the manufacturing-related modules. Another company might consider its HR department's operations so separate from its other operations that it would not integrate its HR functional area. Another company might believe that its internally developed production and logistics software gives it a competitive advantage, and so it would implement the SAP R/3 Financial Accounting and Human Resources modules, but integrate its internally developed production and logistics system into the SAP R/3 system.

Generally, a company's level of data integration is highest when the company uses one vendor to supply all of a company's modules. When a company uses modules from different vendors, additional software programming must be done to get the modules to work together. Frequently, companies integrate different systems using batch data transfer processes that are performed periodically. In this case, the company no longer has accurate data available in real-time across the enterprise. Thus, a company must be sure the decision to use multiple vendors or to maintain a legacy system is based on sound business analysis, not on a resistance to change. Software upgrades of nonintegrated systems are made more problematic because further work must be done to get software from different vendors to interact.

After a company chooses its major modules, it chooses from an assortment (about 8,000) of configuration options. These configuration options allow the company to customize the modules it has chosen to fit the company's needs. For example, in the FI module, a business might need to define limits on the dollar value of business transactions that an employee can process. This is an important consideration in minimizing the risk to fraud and abuse. In configuring the SAP system, the company can define **tolerance groups**, which are specific ranges that define these limits. An example of a tolerance group is shown in Figure 2-6. As part of the configuration process, a company can define any number of tolerance groups with a range of limits and can then assign employees to these tolerance groups.

While SAP has defined the tolerance group methodology as its method for placing limits on an employee, configuration allows the company the flexibility to further tailor this methodology. Let's assume an outdoor clothing company places an order in July for 1,000 ski jackets. The company receives only 995 in the shipment from the manufacturer, which arrives in September. This delivery, although it is short five ski jackets, is close enough to the original order that it is accepted as complete. The difference of the five ski jackets represents the tolerance. By defining the tolerance group to accept a variance of a small percentage of the shipment, the company has determined that it is not worth pursuing the five

FIGURE 2-6 A customization example: tolerance groups to set transaction limits

extra ski jackets. Tolerance could indicate a shortage, as in this example, or an overabundance in an order. Thus, an order of 1,005 ski jackets would also be within the tolerance. Tolerance groups should be defined and documented, in part to deal with fraud issues. If an order is short, is it because of the tolerance or is it because the workers on the loading dock stole five ski jackets?

Features of SAP R/3

Not only was SAP R/3 the first software that could deliver ERP integration, it has other features worthy of note. Its most significant characteristics are its usability by large companies, high cost, automation of data updates, and applicability of best practices, as described below.

The original SAP R/3 system targeted very large companies. Prior to the development of ERP systems, it was assumed these giants could never be integrated because of the sheer amount of computing power required to integrate them. Increased computing speeds, however, meant that large companies in a variety of industries, including manufacturing, gas and oil, airlines, and consulting, could have integrated information systems.

Acquiring a SAP R/3 software system is very expensive. In addition, the software is so powerful, many companies find they must buy new hardware to accommodate the software. For a Fortune 500 company, software, hardware, and consulting costs can easily exceed $100 million. Full implementation of all modules can take years.

The modular design of SAP R/3 is based on business processes, such as sales order handling, materials requirement handling, and employee recruiting. When data are entered into the system, data in all related files in the central database are automatically updated. No human input is required to make the changes.

Before SAP R/3's development, IS people felt that software should be designed to reflect a business' practices. As SAP accumulated experience developing information systems, however, the company began to develop models of how certain industries' business processes should be managed in a way that was compatible with the R/3 system. Thus, R/3's design incorporates **best practices**, which means that R/3 designers choose the best, most efficient ways in which business processes *should* be handled. If a buyer's business practices don't follow SAP R/3's design, then the business must redesign its practices so it can use the software. Prior to ERP systems, companies sought vendors to write software to fit their business processes. With the ERP vendors following their idea of best practices, companies have to fit their business processes to the software's best practices. Although some customization is possible during implementation, companies find they must still change the way they work to fit the software.

ERP FOR MIDSIZE COMPANIES

By 1998, most of the Fortune 500 companies had already installed ERP systems, so ERP vendors refocused their marketing efforts on midsize companies (those with fewer than 1,000 employees).[5] Midsize companies represented a ripe, profitable market. For example, midsize European companies have a total yearly budget of $50 billion for IT expenditures. The American market is even larger.[6] To appeal to this new market, SAP developed Industry Solutions, which are specific, preconfigured versions of R/3 tailored for particular industries such as automotive, banking, chemicals, oil and gas. Because they are smaller and more specific, Industry Solutions cost less than R/3 and can be installed more quickly than the standard R/3 product.

Application hosting, in which a third-party company provides the hardware and software support, rather than an internal IS department, is also making ERP systems like SAP more appealing to midsize companies. SAP is now focusing considerable development efforts on smaller customers with its solutions for small and midsize businesses.

SAP, Oracle, and PeopleSoft have some competition from smaller providers of ERP software. For example, Exact software's ERP package called e-Synergy, has similar functionality to the larger products, with seven modules: Human Resources, Document Managing, Financials, Logistics, CRM (customer relationship management), Procurement, and Project. The company, headquartered in the Netherlands, targets small to midsize companies. Also, in 2000, software giant Microsoft acquired Great Plains, a provider of ERP software. It is unclear whether Microsoft's backing of Great Plains will create competition for the big ERP vendors.

Best of Breed Approaches

ERP software vendors emphasize different strengths within their product offerings. For example, PeopleSoft reputedly offers the strongest Human Resources package, while SAP is renowned for its Production Planning and Materials Management modules. As a result, some companies have opted to integrate modules from different ERP vendors, a design approach called "best of breed." Because the modules originate from different vendors, companies must have middleware written to connect the disparate software.

The best of breed approach may rapidly become obsolete, however. Companies are finding it simpler to keep the systems uniform with the same vendor. Upgrades to the software are easier with a single vendor.

Responses of the Software to the Changing Market

In the mid 1990s, many companies complained about the difficulty of implementing the SAP R/3 system. (SAP faced a cancelled implementation by Dell Computers, a lengthy implementation at Owens-Corning, and a lawsuit by the now-bankrupt FoxMeyer drug company.[7]) In response, SAP developed the Accelerated SAP (ASAP) implementation methodology to ease the implementation process. SAP has continued developing implementation methodology. The latest version, Solution Manager, is designed to greatly speed the implementation process.

SAP continues to extend the capabilities of SAP R/3 with additional, separate products that run on separate hardware and that extract data from the SAP R/3 system. In many cases, these products provide more flexible and powerful versions of tools available in the SAP R/3 system. For example, customers needing more capability and flexibility beyond the standard SD, PP, and FI modules can add the Business Warehouse (BW) product. BW lets the user define unique reporting and analysis methods and integrate information from other systems. The downside of the BW product is that it represents another system to purchase and implement.

The SAP R/3 Enterprise system provides some tools to manage customer interactions and analyze the success of promotional campaigns, but SAP sells a separate program called Customer Relationship Management (CRM), which has extended customer service capabilities. SAP's CRM product is designed to compete with CRM systems from competitors like Siebel.

SAP addresses the issue of Internet-based data exchange with an integration platform called NetWeaver, which allows users to connect SAP products through the Internet.

Like all technology, ERP software and related products are constantly changing. Thus, the challenge for a company is not only to evaluate an ERP vendor's current product offerings, but to also assess its development strategies and product plans.

CHOOSING CONSULTANTS AND VENDORS

Because ERP software packages are so large and complex, one person can't fully understand a single ERP system; it is also impossible for an individual to compare various systems. So, before choosing a software vendor, most companies study their needs and then hire an external team of consultants to help choose the right software vendor(s) and the best approach to implementing ERP. Working as a team, the consultants apply their expertise to selecting an ERP vendor (or vendors) that best meets their customer's needs.

After selecting a vendor, the consultants recommend the modules that are best suited to the company's operations and the configurations within those modules that are most appropriate. This preplanning should involve not only the consultants and a company's IT department, but management as well.

THE SIGNIFICANCE AND BENEFITS OF ERP SOFTWARE AND SYSTEMS

The significance of ERP lies in its many benefits. As was previously mentioned, integrated information systems can lead to more efficient business processes that cost less than those in unintegrated systems. ERP systems have these benefits as well:

- ERP allows easier global integration: Barriers of currency exchange rates, language, and culture can be bridged automatically, so data can be integrated across international borders.
- ERP not only integrates people and data, but it also eliminates updating and repairing many separate computer systems. For example, Boeing had 450 data systems that fed data into its production process. The company now has a single way to record production data.[8]
- ERP allows management to manage operations, not just monitor them. For example, without ERP, getting an answer to "How are we doing?" requires getting data from each business unit and then putting the data together for a comprehensive, integrated picture. The ERP system already has all the data, allowing the manager to focus on improving processes. This focus enhances management of the company as a whole, and makes the organization more adaptable when change is required.

QUESTIONS ABOUT ERP

How Much Does an ERP System Cost?

Cost of an ERP system includes several factors:

- The size of the ERP software, which corresponds to the size of the company it serves
- New hardware that is capable of running complex ERP software
- Consultants' and analysts' fees
- Time for implementation (which causes disruption of business)
- Training (which costs both time and money)

A large company, one with well over 1,000 employees, will likely spend $50 million to $500 million for an ERP system with operations involving multiple countries, currencies, languages, and tax laws. Such an installation might cost as much as $30 million in software license fees, $200 million in consulting fees, additional millions to purchase new hardware, and even more millions to train managers and employees.[9] In addition, full implementation of the new system might take four to six years.

A midsize company (one with fewer than 1,000 employees) might spend $10 million to $20 million in total implementation costs and have its ERP system up and running in about two years.

Should Every Business Buy an ERP Package?

ERP packages imply, by their design, a way of doing business, and they require users to follow that way of doing business. Some of a business's operations, and some segments of its operations, might not be a good match with the constraints inherent in ERP. Therefore, it is imperative for a business to analyze its own business strategy, organization, culture, and operation *before* choosing an ERP approach.

An article in the *Harvard Business Review* provides examples that show the value of planning: "Applied Materials gave up on its system when it found itself overwhelmed by the organization changes involved. Dow Chemical spent seven years and close to half a billion dollars implementing a mainframe-based enterprise system; now it has decided to start over again on a client-server version."[10] In another example, Kmart in 2002 wrote off $130 million because of a failed ERP supply chain project. At the time Kmart was not happy with its existing supply chain software, and it attempted to implement another product too quickly. In addition, the CIO left the company and new management wanted to make some changes.[11]

Sometimes, a company is not ready for ERP. In many cases, ERP implementation difficulties result when management does not fully understand its current business processes and cannot make implementation decisions in a timely manner. An advantage of an ERP system is that it can reduce costs by streamlining business processes. If a company is not prepared to change its business processes to make them more efficient, then it will find itself with a large bill for software and consulting fees with no improvement in organizational performance.

Is ERP Software Inflexible?

While many people claim that ERP systems, especially the SAP R/3 system, are rigid, SAP R/3 does offer numerous configuration options. In addition, programmers can write specific routines for special applications in SAP's internal programming language, called **Advanced Business Application Programming (ABAP)**. The new integration platform, NetWeaver, also offers more flexibility in adding both SAP and non-SAP components. Companies should be careful in how much custom programming they include in their implementations, however; they might find that they have just recreated their existing information systems in a new software package, rather than gaining the benefits of improved, integrated business processes.

Once an ERP system is in place, trying to reconfigure it while retaining data integrity is expensive and time-consuming. That is why thorough pre-implementation planning is so important. It is much easier to customize an ERP program during system configuration and before any data have been stored.

ERP sometimes does not work well for companies that experience rapid growth and change. For example, in 1997 Dell Computer cancelled most of a two-year-old SAP R/3 project when tests showed the software couldn't handle anticipated sales volume.[12] At that time, Dell's operation was growing by $1 billion every six to eight weeks. The company's rapid growth would have meant constant and costly updating of the ERP installation. In addition, the company could not find a single vendor offering all the options it wanted. Instead, Dell implemented flexible middleware software, allowing the company to buy software from a variety of vendors and make changes to its data that would keep up with the company's rapid expansion.[13] Expensive? Yes. The right solution? For Dell, at that time, yes.

What Return Can a Company Expect from Its ERP Investment?

The financial benefits provided by an ERP system can be difficult to calculate because sometimes ERP increases revenue and decreases expenses in intangible ways that are difficult to measure. Also, some changes take place over such a long period of time that they are difficult to track. Finally, the old information system may not be able to provide good data on the performance of the company before the ERP implementation. Still, the return on an ERP investment can be measured and interpreted in many ways:

- Because ERP eliminates redundant effort and duplicated data, it can generate savings in operations expense. One study indicated that 33 percent of companies saw a cost savings in sales order management, and 34 percent of companies said their ERP system significantly reduced their personnel needs.[14]
- Because an ERP system can help produce goods and services more quickly, more sales can be generated every month.
- In some instances, a company that doesn't implement an ERP system might be forced out of business by competitors that have an ERP system—how do you calculate the monetary advantage of remaining in business?
- A smoothly running ERP system can save a company's personnel, suppliers, distributors, and customers much frustration—a benefit that is real but difficult to quantify.
- Because both cost savings and increased revenues occur over many years, it is difficult to put an exact dollar amount accrued from the original ERP investment.
- Because ERP implementations take time, there may be other business factors affecting the company's costs and profitability, making it difficult to isolate the impact of the ERP system alone.
- Because ERP systems provide real-time data, companies can improve external customer communications. Better communication can improve the customer relationship and increase sales.

How Long Does It Take to See a Return on an ERP Investment?

A **return on investment (ROI)** is an assessment of an investment project's value that is calculated by dividing the value of the project's benefits by the value of the project's cost. An ERP system's ROI can be difficult to calculate because of the many intangible costs and benefits previously mentioned. Some companies do not even try to make the calculation, on the grounds that the package is as necessary as having electricity (which is not justified as an investment project). Companies that do make the ROI calculation have seen widely varying results. Some ERP consulting firms refuse to do ERP implementations unless their client company performs an ROI.

Some companies claim that an acceptable ROI begins to accrue almost immediately following the ERP system installation. For example, Pitney Bowes, a $4 billion-per-year manufacturer of fax machines, mailing systems, and copiers, said that ERP rapidly cut its costs of overall operations by 28 percent. In addition, sales representatives can give customers accurate price quotes in 15 minutes rather than several hours, which immediately increased sales nearly 4 percent. (A 4 percent increase doesn't sound like much, but 4 percent of $4 billion is $160 million!) The new system has also increased sales representatives' data entry accuracy by 41 percent, which translates to happy customers who are getting the right products faster.[15]

Although Nestle USA has had problems with its ERP implementation, it estimated a cost savings of $325 million, after spending six years and over $200 million on the implementation.[16]

Toro, a wholesale lawnmower manufacturer, spent $25 million and four years to implement an ERP system. At first, ROI was difficult for Toro to quantify. Then, the emergence of an expanded customer base of national retailers, such as Sears and Home Depot, made it easier to quantify benefits. For example, Toro was able to gain a yearly savings of $10 million inventory reduction—the result of better production, warehousing, and distribution methods.[17]

On the other hand, a few companies claim that high ERP costs adversely affected profits. Pharmaceutical firm FoxMeyer claims that its SAP implementation caused its bankruptcy.

Why Do Some Companies Have More Success with ERP Than Do Others?

Some reports indicate that only a low percentage of companies experience a smooth rollout of their new ERP systems *and* immediately begin receiving the benefits they anticipate. You should put such reports into perspective. *All* kinds of software implementations can suffer from delays, cost overruns, and performance problems, not just ERP projects. Such delays have been a major problem for the IS industry since the early days of business computing. Nevertheless, it is worth thinking specifically about why ERP installation problems can occur.

You can find numerous cases of implementation woes in the news. W.L. Gore, the maker of GoreTex, had some problems implementing its PeopleSoft system for personnel, payroll, and benefits. The manufacturer sued PeopleSoft, Deloitte & Touche LLP, and Deloitte Consulting for incompetence. W.L. Gore blamed the consultants for not understanding the system and leaving its personnel department in a mess. PeopleSoft consultants were brought in to fix the problems, but the fix cost W.L. Gore additional hundreds of thousands of dollars.[18] Hershey foods had a rough rollout of its ERP system in 1999, due to what experts say was the "Big Bang" approach to implementation, in which huge pieces of the system are implemented all at once. Companies rarely use this approach because it is so risky. Hershey lost a large share of the Halloween candy market that year due to ERP problems from this poor implementation.[19]

Usually, a bumpy rollout and low ROI are caused by *people* problems and misguided expectations, not computer malfunctions. For example:

- Some executives blindly hope that new software will cure fundamental business problems that are not curable by any software.
- Some executives and IT managers don't take enough time for a proper analysis during the planning and implementation phase.

- Some executives and IT managers skimp on employee education and training.
- Some companies do not place the ownership or accountability for the implementation project on the personnel who will operate the system. This lack of ownership can lead to a situation where the implementation becomes an IT project rather than a company-wide project.
- Unless a large project such as an ERP installation is promoted from the top down, it is doomed to fail. The top executives must be behind the project 100 percent for it to be successful.
- ERP implementation brings a tremendous amount of change for the users. Managers need to manage that change well so that the implementation goes smoothly.

Many ERP implementation experts stress the importance of proper education and training for both employees and managers. Most people will naturally resist changing the way they've done their jobs. Many analysts have noted that active top management support is crucial for successful implementation.

Some companies willingly part with funds for software and new hardware, but don't properly budget for employee training. ERP software is complex and can be intimidating at first. This fact alone supports the case for adequate training. Typically, ERP training costs $10,000 to $20,000 per employee.[20] Some analysts recommend allocating 11 percent of the project's budget for training.[21] The cost includes training employees on how to use the software to do their job, employees' nonproductive downtime during training, and—very important—educating employees about how the data they control affects the entire business operation.

Nestle has learned many lessons from its implementation of ERP systems. Its six-year, $210 million project was initially headed for failure because Nestle didn't include on the implementation team any employees from the operating groups affected. Employees left the company, morale was down and help desk calls were up. After three years, the ERP implementation was temporarily stopped. Jeri Dunn, the vice president and CIO of Nestle USA learned that the project was not about implementing the software, but about change management. "When you move to SAP, you are changing the way people work... You are challenging their principles, their beliefs and the way they have done things for many, many years," said Dunn.[22] Nestle ultimately reaped benefits from its ERP installation.

Most ERP installations do generate returns. Only 4 percent of managers reported that they had achieved *all* the benefits that they had anticipated from their ERP installation. Thirty-eight percent of managers have achieved the majority of benefits they expected, and 38 percent have achieved about half of the anticipated benefits.[23]

ANOTHER LOOK

ERP, E-Commerce, and Back-Office Systems

The DuPont Chemical company is in the midst of a large SAP implementation. Co-author Ellen Monk interviewed some members of the implementation team (Pat Ryan, Global Process Lead for ATP; Andy Flood, Project Lead DuPont IT for Floor Products; and Omar Boulos, Project Manager, Accenture) about the project.

Why was ERP attractive to the DuPont company?

In the floor products division, we had no global view of sales and revenue. The application architecture in this group consisted of about 500 legacy systems. We couldn't do capacity planning across sites that made the same product. Our customers wanted to order online so we needed a robust back-office system that would serve our e-commerce push. Our philosophy was that ERP was the "price of staying in business." Before the project decision, we did lots of financial analysis that was required by the corporation; for example, benefits over time. The primary savings area was in the supply chain.

Give me a sense of the magnitude of this project.

We are organized in 15 legal entities with strategic business units beneath them. Decisions are made at the strategic business unit level, but the SAP system was implemented globally at the financial, or legal entity, level. We implemented nearly every module of SAP. The rollout was driven by business unit and region. We had to have an interim solution running while implementing the new SAP system to keep the supply chain moving.

The project started in January 2001 and we finished the fourth rollout in August 2004. The first rollout was the flouro chemical division, U.S.-based. The division has four DuPont manufacturing facilities and 12 contract manufacturing facilities. There are 600 end users and the business brings in $700 million.

How did you handle change management and training?

Change management and communications were handled in three steps:

Step 1 – Project team formation: experienced line managers and professionals that were highly respected were made full-time [employees] on the project. The vice president sent out an endorsement of the project to the world. We consider this a very critical step.

Step 2 – Project sponsor: personal trips were made to 15 major plant sites worldwide (Andy visited the plants each four times) to communicate face to face with plant leadership. These trips ensured adequate staffing for the plant and answered any questions the plant personnel might have had.

Step 3 – Global operations director: her view was that on day 1 of start-up if any plant had problems (such as if they couldn't ship material) she blamed the plant manager. So the plant manager very clearly owned the success or failure of the project.

In addition, there were weekly project team letters, two full-time teams on change management, Intranet Web sites, and videotapes with information on the process areas, changes, and benefits.

Training was handled by professional trainers, using detailed documentation. Users had to pass a qualification-based test – which was easy if they attended and participated in

continued

the training sessions. If not, they had to retrain and retest. Training is a journey, not a beginning and end. Users were trained in the entire process and then were able to drill down to their own specific job.

Questions:

1. Why did the DuPont company need to revamp their back-office systems? Do you agree with the quote that ERP is the "price of staying in business"?

2. How did DuPont handle change management with its employees? Can you think of any improvements the company could have made to alleviate user stress?

3. Research change management and systems development on the Internet and report your findings to your instructor.

THE CONTINUING EVOLUTION OF ERP

Understanding the social and business implications of new technologies is not easy. Howard H. Aiken, the inspiration behind the first large-scale digital computer, the Harvard Mach I, predicted in 1947 that only six electronic digital computers would be needed to satisfy the computing needs of the entire United States! Hewlett-Packard passed up the opportunity to market the computer created by Steve Wozniak that became the Apple I. Microsoft founder Bill Gates did not appreciate the importance of the Internet until 1995, by which time Netscape controlled the bulk of the Internet browser market. (Gates, however, did dramatically reshape Microsoft around an Internet strategy by the late 1990s. Its Internet browser is now as commonly used as any other.) Thus, even people who are most knowledgeable about a new technology do not always fully understand its capabilities or how it will change business and society.

ERP systems have been in common use only since the mid-1990s. As this young technology continues to mature, ERP vendors are working to solve adaptability problems that plague customers. Fundamental developments are discussed next.

Additional Capabilities Within ERP

Sales production, data analysis, and Internet connectivity are a few areas where ERP vendors are expanding ERP capabilities. As discussed previously, ERP vendors and other software companies are continuing to develop CRM applications that increase the efficiency of the sales force. For example, if salespeople can target the most profitable customers—actual and potential—they'll have a competitive advantage (you will learn more about CRM in Chapter 3). Other software focuses on detecting changes in customer satisfaction and responding quickly to remedy any problems.

Supply Chain Management (SCM) is one area that can benefit from ERP. SCM applications help to translate customer demand into production plans more efficiently. For example, SAP's Advanced Planner and Optimizer module provides sales representatives with a Global Available-to-Promise (ATP) capability. Global ATP allows the sales representatives to instantly check all available plants and warehouses in the company to find the best option for meeting a customer order in a timely and cost-efficient way. If an office supply company sends in an order for 100 computer printers to a printer manufacturer specifying delivery the following week, the manufacturer will look at its inventory of printers and production of printers. If it has 100 printers in inventory, and 100 printers are to be manufactured the following week, the manufacturer will assign the newly manufactured items to the office supply company. This process is done in ATP, which assigns the future manufacturing run to the office supply company because the printer manufacturer could receive an additional order for 100 printers that needs to be filled today.

ERP developers are also trying to make their existing systems smarter by extending ERP's capabilities into more areas of decision support, management reporting, and data mining. **Data mining** is the statistical and logical analysis of large sets of transaction data, looking for patterns that can aid decision-making. For example, discovering patterns in customer behavior can lead to better marketing efforts. Strategic Enterprise Management (SEM) applications help a company translate corporate-level goals (such as profit and market share targets) into operational decisions (such as production plans and workforce levels).

In addition, ERP vendors continue to improve software and Internet connections that integrate a business' internal operations, while also integrating the business with its dealers, vendors, and customers. Web services is one area that is receiving a lot of attention.

The Internet

The Internet's rapid development since the mid-1990s has been a threat to ERP software developers. ERP software lets users access the company's software and central database through internal connections. Now, users often need to access that central database directly from the Internet. Access through the Internet allows for greater flexibility in work because all one needs is a computer with an Internet connection and a Web browser. This has forced ERP companies to rethink how users get to and use their ERP software. ERP developers have been incorporating Web-based systems with their ERP products. This effort is typified by SAP's NetWeaver. NetWeaver is an application that lets companies add components to their SAP R/3 systems and also lets external partners access certain parts of the company's ERP system. Such open access is a necessity for the success of Internet-based business activities such as **electronic commerce** (or e-commerce, the conduct of business over the Internet). The Internet has become an important way to sell goods and services and will probably become more important in the future. Companies will have a continuing need to take orders electronically and to pass them seamlessly to the company's database. At that point, the ERP system can manage the transaction as if it had come in through a traditional method, such as the telephone.

Trading hubs on the Web integrate companies' ERP systems for exchange of goods and services. Elemica, the chemical industry trading site, includes such companies as Air Products, BASF, BP Chemicals, DuPont, and Procter & Gamble and allows them to conduct business with one another.[24]

In conducting e-commerce, companies put a lot of focus on their sites' Web interfaces, spending great amounts of time and money to ensure that the site works smoothly. A company that intends to sell its products on the Web, however, must still manage its business processes. Some experts have speculated that e-commerce will make ERP systems obsolete, but the disappearance of ERP is not likely to occur. Rather, e-commerce is another activity that ERP systems can help manage. Chapter 8 will elaborate on the importance of the Internet in managing ERP systems.

ANOTHER LOOK

Maximizing Value from an ERP System

A 2002 study by the Accenture Institute for Strategic Change found that companies could not simply implement an ERP system and then forget about it.[25] The implementation of ERP is an ongoing process, not a one-time task. Users have to get to know the system before they can reap full benefits. To achieve real value from the system, Accenture concluded that companies should:

- *Integrate.* To gain the full benefit of ERP systems, they must be integrated throughout the company. Many companies have disparate information systems, resulting either from acquisitions or from legacy systems. Integrating these disparate systems is important. Executives are being told to replace the multiple systems their companies use today with one ERP system that serves the entire company, no matter how diversified or geographically spread out it is.[26]
- *Optimize.* To get further benefit from your ERP system, you need to optimize, or perfect, it. Many companies rushed their ERP implementations to meet the Y2K deadline. Now is the time for management to go back, rethink business processes, and optimize their systems for further savings.
- *Informate.* "Informate" is a word Accenture uses to describe the process of translating data into useable information. An ERP system can provide management with a single, accurate set of data for decision making. How that data is accessed and presented is up to the users of the system. It can take months for people to understand what they can do with the data. Turning that rich data into information that managers can effectively use takes time and work.[27]

Questions:

1. Based on Accenture's findings, explain why ERP systems' implementations are works in progress, compared to traditional systems implementations.
2. Assume you are the chief information officer of a company that implemented an ERP system for Y2K. Your system is running, but you want to achieve more from it. Bearing in mind the three Accenture recommendations, what steps would you take to obtain more value from your system?

Chapter Summary

1. Several factors led to the development of ERP:

   - The speed and power of computing hardware increased exponentially, while cost and size decreased.

   - The early client-server architecture provided the conceptual framework for multiple users sharing common data.

   - Increasingly sophisticated software facilitated integration, especially in two areas: A/F and manufacturing resource planning.

   - The growth of business size, complexity, and competition made business managers demand more efficient and competitive information systems.

2. SAP AG produced a complex, modular ERP program called R/3. The software could integrate a company's entire business by using a common database that linked all operations, allowing real-time data sharing and streamlined operations.

3. SAP R/3 is modular software offering modules for Sales and Distribution, Materials Management, Production Planning, Quality Management, and other areas.

4. ERP software is expensive to purchase, time consuming to implement, and requires significant employee training, but the payoffs can be spectacular. For some companies, however, the ROI may not be immediate or even calculable.

5. Experts anticipate that ERP's future focus will be on managing customer relationships, improving planning and decision making, and linking operations to the Internet.

Key Terms

Advanced Business Application Programming (ABAP)

Asset Management (AM) module

Best practices

Client-server architecture

Controlling (CO) module

Data mining

Database management system (DBMS)

Electronic commerce (e-commerce)

Electronic data interchange (EDI)

Financial Accounting (FI) module

Human Resources (HR) module

Legacy system

Material requirements planning (MRP)

Materials Management (MM) module

Modules

Open architecture

Plant Maintenance (PM) module

Production Planning (PP) module

Project System (PS) module

Quality Management (QM) module

R/3

Return on investment (ROI)

Sales and Distribution (SD) module

Sales forecast

Scalability

Tolerance groups

Workflow (WF) module

Exercises

1. Define Moore's Law and explain why it is significant in the development of ERP.

2. What are the main characteristics of an ERP system? List the benefits and disadvantages of implementing an ERP system.

3. Imagine that you have a job interview with the owner of a surfboard manufacturing company that employs about 300 people. The owner of the company has heard about ERP, but she hasn't done any research on it. She asks you, "Do you think we need an ERP system?" What would you say to her? What questions might you ask to help her think about ERP in her company? List your response and questions. Do research on small ERP installations to back up your response. (*Hint*: Check out InformationWeek.com: Big Ambitions, 8/25/2003.)

4. Much has been written about ERP, both in the popular press and on the Internet. Using library resources or the Internet, report on (1) one company's positive experience with implementing ERP, and (2) one company's disappointing experience.

5. Companies whose common stock is traded on U.S. stock exchanges are required to publish an annual report to stockholders. Obtain the annual reports of SAP, Oracle, and People-Soft for the past three years. Write a report comparing how the companies say their business has changed in that time. Include charts showing changes in important operational and financial results. Also include the take-over bids offered by Oracle for PeopleSoft.

6. Visit the online magazine CIO.com/ERP. Choose an example of an ERP implementation and write a memo to your instructor describing the procedure. Make comments as to areas the company could have improved its implementation.

End Notes

[1] Thomas H. Davenport, "Putting the Enterprise into the Enterprise System," *Harvard Business Review* (July-August 1998): 121–31.

[2] http://www.peoplesoft.com/corp/en/case_studies (2002).

[3] Robert Westervelt, "U.S. software sales boost SAP earnings, says CEO," SearchSAP.com (April 22, 2004).

[4] Gerd Meissner, *Inside the Secret Software Power*, McGraw-Hill (1997).

[5] Kuldeep Kumar and Jos van Hillegersberg, "ERP Experiences and Evolution," *Communications of the ACM* 43, No. 4 (April 2000): 23–26.

[6] Yvonne van Everdingen, Jos van Hillegersberg, and Eric Waarts, "ERP Adoption by European Midsize Companies," *Communications of the ACM* 43, No. 4 (April 2000): 27–31.

[7] Davenport, "Putting the Enterprise into the Enterprise System," 121–31.

[8] Polly Schneider, "Human Touch Sorely Needed in ERP," *CIO* (online) (March 2, 1999).

[9] David Kirkpatrick, "The E-Ware War," *Fortune* (December 7, 1998): 102–12.

[10] Davenport, "Putting the Enterprise into the Enterprise System," 121–31.

[11] Carol Sliwa, "IT difficulties help take Kmart Down," *Computerworld*, (January 28, 2002).

[12] Joseph B. White, Don Clark, and Silvia Ascarelli, "Program of Pain," *Wall Street Journal* (March 14, 1997).

[13] Derek Slater, "How to Choose the Right ERP Software Package," *CIO* (online) (February 1999).

[14] Tom Stein, "Making ERP Add Up," *Information Week.com* (May 24, 1999).

[15] Stein, "Making ERP Add Up."

[16] "Nestle's ERP Odyssey," *CIO* (May 15, 2002).

[17] Stein, "Making ERP Add Up."

[18] Elizabeth MacDonald, "W.L. Gore Alleges PeopleSoft, Deloitte Botched a Costly Software Installation," *Wall Street Journal* (November 2, 1999).

[19] Emily Nelson and Evan Ramstad, "Hershey's Biggest Dud is its new Computer System," *Wall Street Journal* (November 4, 1999).

[20] Richard Stooker, "SAP: A Hot Skill—The Current #1 Resource Planning Software Package," *Inforing Press* (online) (2000).

[21] Malcolm Wheatley, "ERP Training Stinks," *CIO* (online) (June 1, 2000).

[22] "Nestle's ERP Odyssey."

[23] "By the Numbers," *CIO* (November 1, 2003).

[24] "Connecting the Chemical Industry," *Chemical Week* (September 25, 2002).

[25] Davenport, "Putting the Enterprise into the Enterprise System," 121–31.

[26] "Extreme ERP Makeover," *CIO* (November 15, 2003).

[27] "The Information Cannot Speak for Itself," *IntelligentEnterprise.com* (July 10, 2004).

CHAPTER **3**

MARKETING INFORMATION SYSTEMS AND THE SALES ORDER PROCESS

LEARNING OBJECTIVES

After completing this chapter, you should be able to:

- Describe the unintegrated sales processes of the fictitious Fitter Snacker company.
- Explain why unintegrated Sales and Marketing information systems lead to company-wide inefficiency, higher costs, lost profits, and customer dissatisfaction.
- Discuss sales and distribution in the SAP R/3 system, and explain how integrated data sharing increases company-wide efficiency.
- Describe how SAP R/3 processes a standard sales order.
- Describe the benefits of Customer Relationship Management software, a useful extension of ERP software.

INTRODUCTION

In this chapter, you will begin reading about the operations of Fitter Snacker (FS), a fictitious company that

makes healthy snack bars and does not have an integrated information system. Throughout the remainder

of this book, we will use FS to illustrate information systems concepts in general and ERP concepts in

particular.

As is the case in many other companies, Marketing and Sales (M/S) is the focal point of many of FS's activities. Why? Because Marketing and Sales is responsible for selling the company's product. As a result, Marketing personnel often guide the company's key strategies and tactics. Marketing personnel in most companies make the following kinds of decisions:

- What products should we produce?

- How much of each product should we produce?

- How are our products best promoted and advertised?

- How should our products be distributed for maximum customer satisfaction?

- What price should we charge for our products?

On a day-to-day basis, M/S is involved in generating key transaction data (including data for recording sales), creating customers' bills, and allocating credit to customers. FS's M/S information systems are not well integrated with the company's other information systems. This results in inefficient use of transaction data, as you will see in this chapter. You'll also see how FS's M/S information systems could be improved using ERP. We begin by looking at an overview of the company's operations. As you have learned in previous chapters, the availability of a common database to all modules is one of the advantages of having an integrated information system, because the data in the system are consistent. Integration can lead to problems, however; if the data are not correct, the error will carry over into all modules.

OVERVIEW OF FITTER SNACKER

Fitter Snacker manufactures and sells two types of nutritious snack bars: NRG-A and NRG-B. NRG-A touts "advanced energy." NRG-B boasts "body building proteins." Each bar contains the following ingredients:

- Vitamins and Minerals: important vitamins and minerals
- Dry Base Mixture: oats, wheat germ, protein powder, and spices
- Wet Base Mixture: honey and canola oil

Each type of bar contains additional ingredients that differentiate the bars: NRG-A contains carob chips and raisins, while NRG-B contains hazelnuts and dates.

Fitter Snacker has organized its sales force into two groups, known as divisions: the Wholesale Division and the Direct Sales Division. The Wholesale Division sells to middlemen who distribute the bars to small shops, vending machine operators, and health food stores. The Direct Sales Division sells directly to large grocery stores, sporting goods stores, and other large chain stores. Each division has an organizational structure that interacts with FS's other functional areas, such as Accounting and Supply Chain Management. The two divisions operate separately from one another, in effect breaking the M/S functional area into two pieces.

The two sales divisions differ primarily in terms of quantities of orders and pricing terms. The Direct Sales Division offers customers volume discounts to encourage larger sales orders, which are more efficient to process because each order generates costs related to the paperwork, shipping, and handling of the order. A large order of 500 cases of snack bars might generate $5,000 worth of profit minus one set of handling costs. A small order of 10 cases of bars might generate only $100 with the same handling costs. The Wholesale Division charges customers a lower fixed price because the orders are usually large (otherwise, the orders would be handled by the Direct Sales Division). In both divisions, customers are sent an invoice and offered a 2 percent discount if they pay within 10 days; otherwise, the total balance is due in 30 days (2–10/net 30).

In addition to selling snack bars under the Fitter Snacker brand name, the company also packages the bars in store-brand wrappers for some chain stores.

PROBLEMS WITH FITTER SNACKER'S SALES PROCESS

Many of Fitter Snacker's sales orders have some sort of problem, such as incorrect pricing, excessive calls to the customer for information, delays in processing orders, missed delivery dates, and so on. These problems are a result of FS having separate information systems throughout the company for three functional areas: the sales order system, the warehouse system, and the accounting system. Information in each system is shared through periodic file transfers (sales order system to accounting system) and manually by paper printout (credit status from Accounting to sales clerks). The high number of transactions that are handled manually creates many opportunities for data entry errors. Further, all the information stored in the three systems is not available in real time, resulting in incorrect prices and credit information.

In each sales division, Fitter Snacker has four salespeople who work on the road, plus two sales clerks who work in the sales office. Salespeople work on commission and have some leeway in offering customers "discretionary discounts" to make a sale. The entire sales process involves a series of steps that require coordination between Sales, Warehouse, Accounting, and Receiving, as shown in Figure 3-1. (Notice that Manufacturing is usually not directly involved in the sales process because goods are shipped from the warehouse.)

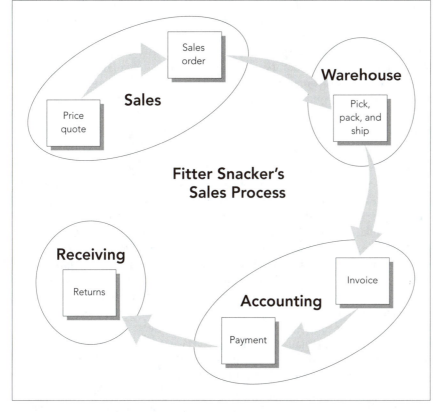

FIGURE 3-1 The sales process

Sales Quotations and Orders

Giving a customer a price quotation and then taking the customer's order should be a straightforward process, but at FS it is not. For a new customer, the sales process begins with a sales call, during which the salesperson either telephones the customer or visits in person. At the end of the sales call, the salesperson prepares a handwritten quotation on a three-part form. The form generates two copies; the top sheet goes to the customer, the middle copy is first faxed and then mailed to the sales office, and the salesperson keeps the bottom copy for his records. The quotation form has an 800 number that the customer can call to place an order.

A number of problems can occur with this process:

- The salesperson might make an arithmetic error in the sales quotation. For example, a salesperson in the Direct Sales Division might offer both a quantity discount plus a discretionary discount. If the salesperson isn't careful, the two discounts combined might be so deep that the company receives little or no profit.

- The salesperson faxes a copy of the sales quotation to the sales office, but sometimes the customer calls to place an order before the fax is transmitted. The in-office clerk has no knowledge of the terms of the sale (which are outlined on the quotation) and must ask the customer to repeat the information. On the other hand, even if the quotation has been faxed, the data might not have been entered into the customer database, and still the customer must repeat the order information, much to his annoyance. This situation can also lead to a duplicate order.
- The fax received by the sales office is a fax of a copy of a handwritten form, and might not be legible.

When customers place an order, they usually ask for a delivery date. To get a shipping date, the in-office clerk must contact the warehouse supervisor and ask whether the customer's order can be shipped from inventory, or whether shipping will be delayed until a future production run is delivered to the warehouse. Because she's too busy to make an inventory count, total all orders waiting to be filled, and find out how many orders are in process in the sales office, the warehouse supervisor can only estimate the shipping date.

Once the in-office sales clerk has the warehouse supervisor's estimated shipping date, she determines the shipping method for the order and how long delivery will take. Next, the clerk checks the customer's credit status. For new customers, the clerk fills out a paper credit-check form that includes basic customer data and the amount of the order. The form is given to Accounting; the accountants perform the credit check and then return the credit-check form showing the customer's credit limit. If the credit limit is below the amount of the purchase, assuming there are no other orders outstanding, the clerk calls the customer to determine what action the customer wants to take (reduce the amount of the order, prepay, or dispute the amount of credit granted). If the order is from an existing customer, then the clerk checks a paper report from Accounting that shows the customer's current balance, credit limit, and available balance. Because the report is generated weekly, it might not reflect a customer's most recent payments.

The sales clerk enters the customer's order into the current order entry system. The computer program performs four important tasks. First, it stores the customer's order data, which are used later to analyze sales performance at the division level. Second, it prints out a packing list and shipping labels for the warehouse to use to pick, pack, and ship the customer's order. Third, it produces a data file for the Accounting department to use for preparing invoices (the file is also used for financial, tax, and managerial accounting, which is discussed in Chapter 5). And fourth, the data file is copied to a disk and entered into a PC-based accounting software program on Mondays, Wednesdays, and Fridays.

Order Filling

Fitter Snacker's process for filling an order is no more efficient than its sales order process. Packing lists and shipping labels are printed twice a day—at noon and at the end of the day. These are hand-carried to the warehouse, where they are hand-sorted into small orders and large orders. Fitter Snacker bars are packed 24 to a display box. The display boxes have promotional printing and are designed to serve as a display case. FS packs 12 display boxes together to form a standard shipping case. The warehouse stores both display boxes and shipping cases. The Production department produces and wraps the bars and packs them in display boxes. Depending on the inventory levels in the warehouse, Production personnel might transfer the display boxes directly to the warehouse, or they might pack the display boxes into shipping cases.

Display boxes and shipping cases are organized in the warehouse by label type (FS brand and store brand). Small orders consist of less than a full shipping case. For small orders, the order picker goes to the warehouse with a handcart and pulls the number of display boxes listed on the packing list. If there are not enough display boxes in the warehouse to fill the order, the picker might break open a shipping case to get the required number of display boxes. If he does this, he is supposed to advise the warehouse supervisor so she can update the inventory records, but sometimes this step is forgotten.

The picker then brings the display boxes back to the order packing area. Here, small orders are packed into a labeled box, with the packing list enclosed, and then they are prepared for shipping by a small package shipper.

Large orders consist of one or more shipping cases. For these orders, the picker uses a forklift to move the appropriate number of shipping cases to the large-order packing area. The cases are labeled for shipping, loaded on a pallet, and attached to the pallet using shrink-wrap plastic for protection. These pallets are shipped either by one of FS's two delivery trucks or by a less-than-truckload (LTL) common carrier.

Inventory levels are managed by a PC database program in the warehouse. Inventory levels are adjusted on a daily basis, using Production records (showing what has been added to the warehouse), packing lists (showing what has been shipped from the warehouse), and any additional sources of data (such as shipping cases that have been opened to pull display boxes). Each month the warehouse staff conducts a physical inventory count to compare the actual inventory on hand with what the inventory records in the PC database show. Fitter Snacker's monthly inventory counts show that inventory records are more than 95 percent accurate. Ninety-five percent accuracy doesn't sound too bad, but FS still has problems filling orders; these are described next.

Because snack bars are somewhat perishable, FS keeps inventory levels fairly low, and inventory levels change rapidly during the day. As a result, pickers might go to the shelves to pick an order and discover that there are not enough of the desired type of snack bars available to fill the order. The picker then has to decide from among several courses of action. There might be more of that type of bar in the production area (on the way to the warehouse) that can be used to complete the order. For an important customer, the wrappers and display box on the production line might be changed to the customer's brand to produce enough bars to complete the order. In other situations, the customer may want a partial shipment of whatever is on hand, with the rest shipped when it becomes available, which is known as a backorder. Or, the customer might prefer a partial shipment of the goods on hand and cancel the balance of the order, placing a new order later. Or, the customer may have enough inventory to wait until the whole order can be shipped, thus saving on delivery charges.

To determine what to do in this situation, the order picker might have conversations with the warehouse supervisor, production supervisor, and sales clerks. Whatever the final decision, the warehouse supervisor has to contact the sales clerk so she can notify the customer (which doesn't always happen when things are busy) and the Accounting department so they can change the invoice.

Accounting and Invoicing

Invoicing the customer is problematic as well. Three times a week, sales clerks send the Accounting department the disk containing the sales order data for customer invoices. The Accounting department loads the data into the PC-based accounting program; then clerks manually make adjustments for partial shipments and any other changes that occurred during the order process. Sometimes order corrections are delayed and don't catch up to the invoicing process, and then invoice delays or inaccurate invoices occur. If the completed invoice is waiting to be mailed when the warehouse notifies Accounting of a partial shipment, then a new invoice must be prepared. In any case, an invoice is eventually sent to the customer, separate from the shipment.

Payment and Returns

Fitter Snacker's procedure for processing payments can yield frustrating results for customers. Almost all customers pay the invoice within 10 days to receive the 2 percent discount. Even though FS provides customers with two invoice copies, many customers don't return a copy of the invoice with their payment, as instructed. Errors result if the correct customer's account isn't credited. In addition, if any errors occurred in the sales process—from the original quotation to the order being keyed into the sales order program—the customer will receive an incorrect invoice.

FS's returns processing is also flawed. Because FS's snack bars contain no preservatives, they have a relatively short shelf life. Thus, FS has a policy of crediting customer accounts for returned snack bars that have exceeded their "sell by" date (this is a generous policy, because it is impossible to know who—FS or the customer—is responsible for the bars not selling before their expiration dates). FS also gives credit for damaged or defective cases. Customers are supposed to call FS to get a returned material authorization (RMA) number to simplify the crediting process. When cases are returned to FS, the Receiving department completes a returned material sheet, listing the returning company's name, the materials returned, and the RMA number. Many times, customers do not call for the RMA number or fail to include it with their returned material. This makes it more difficult for the Accounting department to credit the appropriate account. Poor penmanship on the returned material sheet also creates problems for Accounting.

When accounts become past due, FS sends the customer a dunning letter. The letter informs the customer the account is past due and requests payment—if payment hasn't already been sent. As the account gets more delinquent, the dunning letters usually get more direct and threatening. If a customer's account is not properly credited, the customer may receive a dunning letter in error. Or, the customer might be called mistakenly about exceeding their credit limit after placing a new order. Both situations damage goodwill with both new and repeat customers.

In the following sections, you learn how an ERP system can improve the sales process for Fitter Snacker.

SALES AND DISTRIBUTION IN ERP

An ERP system can improve the sales order process in several ways. Because ERP systems use a common database, they can minimize data entry errors and provide accurate information in real time to all users. An ERP system can also track all transactions (such as invoices, packing lists, RMA numbers, and payments) involved in the sales order.

Let's look at how one ERP system, SAP's R/3 and its Sales and Distribution module, manages the sales order process. Other ERP software handles the process in a similar fashion. In SAP R/3, important transactions and events are assigned a number for record-keeping purposes. The electronic evidence of a transaction in SAP R/3 is called a "document."

SAP R/3 Sales and Distribution module of the SAP R/3 system treats the sales order process as a cycle of events. SAP R/3 defines up to six events for any sale:

- Pre-sales activities
- Sales order processing
- Inventory sourcing
- Delivery
- Billing
- Payment

Pre-Sales Activities

The first step in the SAP R/3 sales and distribution process is Pre-Sales Activities. At this phase, customers can get pricing information about the company's products. This information can be provided either through an inquiry or a price quotation. Both methods provide pricing information to the potential customer. The difference between an inquiry and a quotation is that a written quotation is a binding document; the seller guarantees the buyer that, for some specified period of time, he can buy the product at the quoted price. The inquiry is simply a statement of prices with no guarantee implied.

Pre-sales activities also include marketing activities such as tracking customer contacts, including sales calls, visits, and mailings. The company can maintain data about customers and generate mailing lists based on specific customer characteristics, which enhances targeted marketing efforts.

Sales Order Processing

In the SAP R/3 system, Sales Order Processing is the series of activities that must take place to record a sales order. The sales order can start from a previous quotation or inquiry generated in the pre-sales step. Any information that was collected from the customer to support the quotation (contact name, address, phone number) is immediately included in the sales order.

Some of the more critical steps in Sales Order Processing are recording the items to be purchased, determining the selling price, and recording the order quantities. Users can define various pricing alternatives in the SAP R/3 system. For example, you can use product-specific pricing, such as quantity discounts, or you can define discounts that depend on both the product and a particular customer. Configuring a complex pricing scheme requires a significant amount of programming work, but once you set up the system, it will automatically calculate the correct price for each customer, eliminating many problems that FS experiences.

During Sales Order Processing, the SAP R/3 system checks the Accounts Receivable tables in the SAP R/3 database to confirm the customer's available credit. SAP R/3 adds the value of the order to the credit balance, and then compares the result to the customer's credit limit (also available in the database). If the customer has sufficient credit available, the order is completed. If not, the SAP R/3 system prompts sales personnel to either reject the order, call the customer to check on recent payments, or contact Accounting to discuss any extenuating circumstances.

Inventory Sourcing

When recording an order, the SAP R/3 system checks the company's inventory records and the production planning records to see whether the requested material is available and can be delivered on the date the customer desires. This Available-to-Promise (ATP) check includes the expected shipping time, taking into account weekends and holidays. Fitter Snacker's current system does not provide a good method for checking inventory availability. In the SAP R/3 system, availability is automatically checked and the system can recommend an increase in planned production if a shortfall is expected. SAP also keeps a record of all open orders, so even if the product has not been shipped the system will reserve the product on the shelf and make it unavailable to other customers.

Delivery

In the SAP R/3 system, the word **delivery** means releasing the documents that the warehouse uses to pick, pack, and ship orders, rather than the traditional use of the term to describe the transferal of goods. The delivery process allows deliveries to be created so that the warehouse and shipping activities are carried out efficiently (for example, combining similar orders for picking, or grouping orders based on how and where they are to be shipped).

Once the documents are created by the system for picking, packing, and shipping, the documents are transferred to the Materials Management module, where the warehouse activities of picking, packing, and shipping are carried out.

Billing

Next, the SAP R/3 system creates an invoice by copying the sales order data into the invoice document. This document can be printed and mailed, faxed, or transmitted electronically to the customer. Accounting records are also updated at this point. To record the sale, Fitter Snacker's Accounts Receivable is debited (increased) and Sales is credited. Thus, SAP R/3 updates the accounting records automatically.

Payment

When the customer sends in a payment (physically or electronically), it is again processed by the SAP R/3 system. Cash is debited and the customer's account is credited (reduced). Notice that the timely recording of this transaction has an effect on the timeliness and accuracy of any subsequent credit checks for the customer. Fitter Snacker has had a problem with getting accurate credit checks, frequently blocking orders for companies that are within their credit limit, while granting credit to other companies beyond what is advisable.

A STANDARD ORDER IN SAP R/3

Now we will take a look at how Fitter Snacker's sales order process would work with an SAP R/3 system in place. You will learn how this ERP system would make FS's sales order process work more accurately and efficiently. Notice that ERP allows business processes to cut across functional area lines.

Taking an Order in SAP's R/3

Figure 3-2 shows an order entry screen in SAP R/3's 4.7 Enterprise system, the version of the software released in 2003. The important fields in this screen are summarized in Figure 3-3.

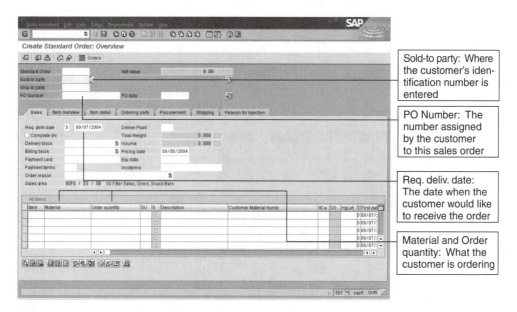

FIGURE 3-2 SAP R/3 order entry screen

In SAP R/3, a sales order clerk must enter code numbers for customers' names and the inventory sold, rather than using customers' names and inventory item names. Because more than one customer might have the same name, a unique number is assigned by the company to each customer in the database. This number acts as the primary identifier for the customer. The same logic applies to distinguishing one inventory item from another. In database terminology, such codes are called key fields. While it sounds like the sales order clerk must remember a lot of code numbers to use the SAP R/3 system, this is not the case. For most data entry fields, the SAP R/3 system determines whether an entry is valid.

Data Entry Field	Explanation
Sold-to party	Identification Number assigned to customer
P.O. Number	The number assigned by the customer to the sales transaction. This is different from the sales order number assigned by the Seller (using SAP R/3) to the sales transaction. In a paper process, the purchase order number is usually a sequential number pre-printed on the purchase order form.
Req. deliv. date	The delivery date for the order requested by the customer. The SAP R/3 system will evaluate the ability to meet this date and suggest alternatives, if necessary.
Material	The identification number assigned in the SAP system to the item requested by the customer.
Order quantity	The number of units of the material the customer is requesting.

FIGURE 3-3   Data entry fields in the order entry screen

Figure 3-4 shows how SAP R/3 lists the 34 sales order types that are included in the system as predefined order types. The sales order clerk can choose the correct type from among these.

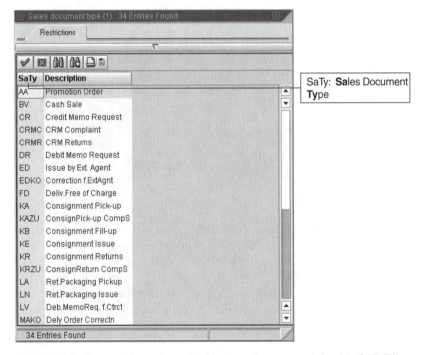

FIGURE 3-4   Some of the sales order (document) types predefined in SAP R/3

When a sales order clerk has to find a customer in the SAP R/3 system, he or she can click the sold-to field to display a search icon and then click the search icon to open the sophisticated search window shown in Figure 3-5. With this search window, the clerk can search for a customer using different criteria (for example, part of the customer's address such as city) to narrow the list produced by the search. Conducting a search using the criteria of distribution channel (specifying direct distribution, shown as DI in the figure), and sales division (specifying snack bar division, shown as SB in the figure) produces a list of customers (Figure 3-6) from which the clerk can pick the correct customer.

Clicking the Sold-to party field produces a search icon. Clicking the search icon calls up a search window with numerous search options.

FIGURE 3-5 Search screen for customers

PostalCode	City	Name 1	Customer
48152	LIVONIA	HEALTH EXPRESS	201
49008	KALAMAZOO	WEST HILLS ATHLETIC CL	1
49544	GRAND RAPIDS	MEIJER, INC.	301
49931	HOUGHTON	HOUGHTON HEALTH CLUB	101

4 Entries Found

FIGURE 3-6 Results of customer search

In the SAP R/3 system, a range of information is stored about each customer in multiple tables. These data are referred to as **customer master data**. Master data are data that remain fairly stable, such as customer name and address data. Master data are maintained in the central database, and are available to all SAP R/3 modules, including SD (Sales and Distribution), FI (Financial), and CO (Controlling).

Information about materials is in tables collectively called **material master data**, which are used by the MM (Materials Management—purchasing and warehousing) and PP (Production Planning) modules in addition to the SD module.

The SAP R/3 system allows the user to define various ways to group customers and salespeople. These groupings are called **organizational structures**. One important organizational structure for Fitter Snacker is the Distribution Channel. With SAP R/3, the Distribution Channel allows the user to define different ways for materials to be sold and distributed to the customer. It allows for different types of relationships with customers and allows the company to specify aspects of the relationship, such as pricing, delivery method, and minimum order quantities. Defining a Wholesale Distribution Channel and a Direct Sales Distribution Channel for FS would help to ensure that customers' orders are correctly priced.

Figure 3-7 shows a completed sales order screen for an order from West Hills Athletic Club for 10 cases of NRG-A bars and 10 cases of NRG-B bars.

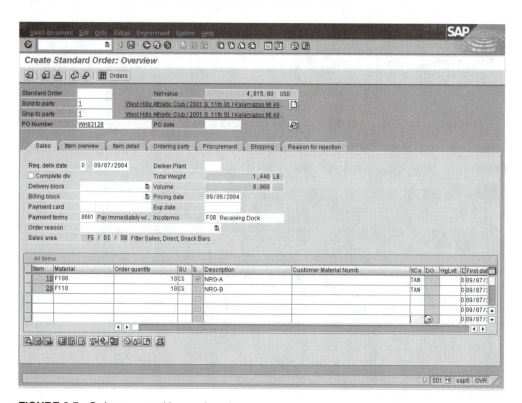

FIGURE 3-7 Order screen with complete data

West Hills's purchase order (PO) number for this sales transaction is WH83128. This is a unique number provided by the customer (West Hills) that allows it to track orders with its suppliers. If the customer has a question about an order, it will reference that PO number in its inquiry. Because the SAP R/3 system records the customer's PO number, FS can look up the status of the order using the customer's PO number in addition to the sales order number that SAP will assign to this transaction.

Notice also that the name of the customer (West Hills Athletic Club) and the names of the products (NRG-A and NRG-B) are shown, even though they were not entered directly during the order process. How did they get on the screen at this point? Once the company code is entered, the clerk can easily request the SAP R/3 system to search the database and accesses all company information needed to complete the order. Thus, the SAP R/3 system simplifies the data entry tasks, reducing data entry time and the possibility of error.

When the SAP R/3 system is instructed to save a sales order, it performs inventory sourcing, meaning it performs checks to ensure that the customer's sales order can be delivered on the requested delivery date. Remember that currently when an FS customer wants to know when an order can be delivered, the sales clerk has to make a series of phone calls. In the SAP R/3 system, this checking is done automatically. It includes checks on both inventory and production, and it includes the time required for shipping. When the requested delivery date cannot be met, the SAP R/3 system automatically proposes alternatives to the sales order clerk. For example, Figure 3-8 shows what the sales order clerk will see if only five cases of NRG-A bars are available by the requested delivery date. The SAP R/3 system provides the sales order clerk with three alternatives:

1. Reduce the order quantity of NRG-A cases to five, which can be delivered by the requested delivery date.
2. Delay shipment of the NRG-A bars by three days, when 10 cases are available.
3. Ship five cases by the requested delivery data, and five cases at the later date.

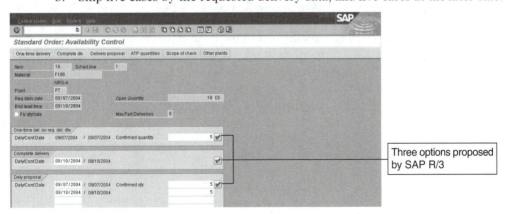

FIGURE 3-8 Order proposals

While the best situation would be to meet the customer's request, when that is not possible the integrated SAP R/3 system can provide the clerk with accurate information about the ability to meet the customer's requirements. With Fitter Snacker's current unintegrated systems, it could take days for the customer to find out that its requested delivery date cannot be met.

When a sales order is saved, the SAP R/3 system assigns a document number to the sales order transaction. In the SAP R/3 system, a document is an electronic record of a business transaction, and a document number is created for each business transaction. When the sales order is ready to be processed by the warehouse, a delivery document will be created with its own unique document number, which the system will relate to the sales order document. Finally, when the bill (invoice) is prepared for the customer, the bill's unique number (called the invoice number) will be created and related to all the other numbers associated with the sales order.

The SAP R/3 system has a mechanism for keeping track of the document numbers for the sales order so that employees can track the status of an order while it is in process or research it after shipping. The linked set of document numbers related to an order, the **audit trail**, is called a **document flow**. Figure 3-9 shows the document flow for a completed order. If an order includes partial shipments, partial payments, and returned material credits, the document flow can become quite complex. Without an integrated information system, the audit trail can be hard to establish, especially if many paper documents are involved. With an integrated system such as SAP R/3, they are all linked together electronically. Not only does the document flow show all documents related to a sales order, but the user can look at the details of each document simply by double-clicking a line in the document flow. For example, if West Hills Athletic Club chose to take delivery of five cases of NRG-A bars on its requested delivery date and five cases when they became available, a sales order clerk can easily check on the status of the five delayed cases using document flow. He or she can search the SAP system to find the original sales order by a number of methods (open orders for West Hills, the PO number that West Hills used, the sales order number assigned by the SAP R/3 system, and so on), and then review the delivery document for the delayed cases. As another example, a customer may have a question about an invoice it received. The SAP R/3 system can be searched for the invoice; the document flow will show all activity that led to the invoice. With an unintegrated information system, researching the invoice would require checking more than one information system and perhaps paper records.

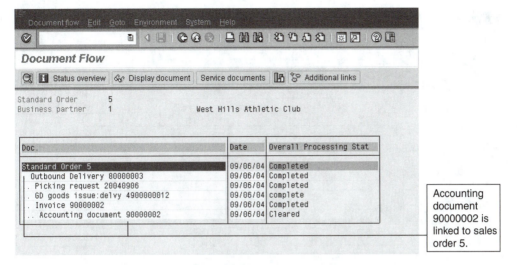

FIGURE 3-9   The Document Flow tool, which links sales order documents

Discount Pricing in SAP R/3

When a company installs an ERP system, they can configure it for a number of pricing strategies. For example, various kinds of discounts can be allowed (per item, on all items, based on unit price, based on total value, with or without shipping charges and taxes, by individual customer, by a class of customer, and so on). In addition, the system can enforce limits on the size of discounts to keep salespeople from offering unprofitable or unapproved discounts.

Suppose a salesperson wants to give a certain customer a 10 percent discretionary discount on NRG-A bars. But is the salesperson allowed to discount those bars? Is that discount appropriate for that customer? If the discount is allowable, will that discount be so deep that the sale will be unprofitable for FS? An ERP system automatically answers these questions.

Pricing, the process of determining how much to charge a particular customer, can be very complex. To accommodate the various ways that companies have developed to offer price discounts, SAP has developed a control mechanism it calls the **condition technique**. While detailed discussion of the condition technique is beyond the scope of this text, Figure 3-10 shows how a discount is automatically applied in the SAP R/3 system.

The screen shows the price that is being offered to West Hills Athletic Club for 10 cases of NRG-A bars. The base price is $240/case, and West Hills is being given a 10 percent discount automatically.

The screen in Figure 3-11 shows that the SAP R/3 system has been configured to give West Hills Athletic Club a 5 percent discount for ordering over $1000 worth of a particular type of snack bar, and a 10 percent discount if the order for a particular type of snack bar is over $1,500 in value.

If you look again at Figure 3-10, you can see that the SAP R/3 system automatically applied the 10 percent discount, so that the net price for 10 cases of NRG-A bars is $2,160. Figure 3-10 also shows that Fitter Snacker will still make a profit on this order, because the cost to produce 10 cases of NRG-A bars is $1,992.

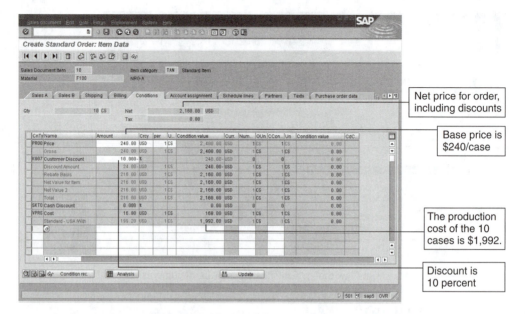

Net price for order, including discounts

Base price is $240/case

The production cost of the 10 cases is $1,992.

Discount is 10 percent

FIGURE 3-10 Pricing conditions for sales order

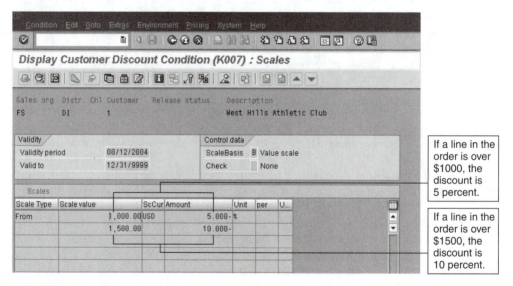

If a line in the order is over $1000, the discount is 5 percent.

If a line in the order is over $1500, the discount is 10 percent.

FIGURE 3-11 West Hills Athletic Club price discount

Integration of Sales and Accounting

In the current FS system, sales records are not integrated with the company's accounting records, so Accounting information is not always up-to-date. By contrast, ERP systems integrate Accounting with all business processes, so that when a sales order is recorded, the related accounting data is updated automatically.

In the document flow shown in Figure 3-9, Accounting document 90000002 is part of the sales order process. The clerk can select this line, then click the Display document button and get the accounting detail shown in Figure 3-12.

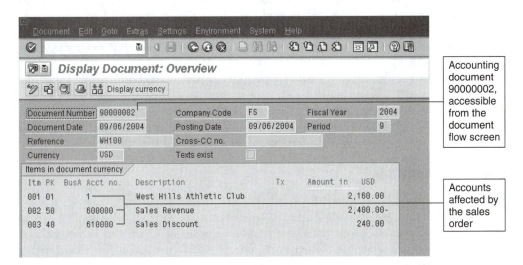

FIGURE 3-12 Accounting detail for the West Hills sales order

This document shows that as a result of this sales order, West Hills Athletic Club has paid Fitter Snacker $2,160. The Sales Revenue account (account number 600000) records sales revenue of $2,400 dollars, which represents the list price of the snack bars West Hills Athletic Club purchased. The 10 percent discount of $240 is recorded in the sales discount account (account number 610000). Because the accounting documents are created automatically with the sales order, the Accounting department is using the same data as Sales, which results in up-to-date and accurate information. In addition, if Marketing needs a report on the size of discounts currently being offered, a report of that data can be easily generated, and it will be up-to-date and accurate.

Now that we have looked at how an ERP system can help a company's sales order process, let's look at how data obtained during that process might be further used to help Fitter Snacker build long-term business relationships with its customers and improve business performance.

CUSTOMER RELATIONSHIP MANAGEMENT

Companies without a good connection between their workers and their customers run the risk of losing business. Take Fitter Snacker, for example. Let's say a salesperson calls on a good customer, Health Express. The salesperson offers Health Express deep discounts for buying a certain number of NRG bars. At the same time, the marketing department is running a sale on NRG bars. Marketing sends Health Express a flyer advertising the discounted price which is less than what the salesperson just offered in person. Meanwhile, the vice president of Fitter Snacker plays golf with the CEO of Health Express and offers yet another discount. The connection or relationship with the customer is confused. **Customer relationship management (CRM) software** can help companies streamline their interactions with customers.

Companies with an ERP system have an added benefit beyond systems integration: complete and vast quantities of data available for analysis. By adding other software tools to their ERP system, companies can extend the capabilities of their ERP system, thus increasing its value. Many ERP vendors—and non-ERP software companies—provide CRM software. Siebel led the CRM software market until 2004, when its market dominance was overtaken by SAP.[1]

With the goal of providing "a single face to the customer," the basic principle behind CRM is that any employee in contact with the customer should have access to all information about past interactions with the customer. In a traditional sales organization, information may only be available to one individual and not shared with the organization. If that individual leaves the company, then the information possessed by that individual is lost.

Core CRM Activities

In general, all CRM software supports the following activities and tools:

- *One-to-One Marketing*. Once a customer is categorized, the company can tailor products, promotions, and pricing accordingly. Customers can be offered products related to what they are now buying (cross-selling) or offered higher-margin products in the same lines (upselling).
- *Sales Force Automation (SFA)*. Occurrences of customer contacts are logged in the company's database. SFA software can automatically route customers who contact the company to a sales representative. Companies can use SFA software to forecast customer needs, based on the customer's history and transactions, and to alert sales representatives accordingly. Sometimes this software is called "lead management software" because a transaction can be tracked from the initial lead to post-sale follow-up.
- *Sales Campaign Management*. This software lets a company organize a marketing campaign and compile its results.
- *Marketing Encyclopedias*. This software serves as a database of promotional literature about products. The material can be routed to sales representatives or customers as needed.
- *Call Center Automation*. When customers call a company to get assistance with a company's products, representatives can query a knowledge management database containing information about the product. Some knowledge management software accepts queries in natural language. If the company must develop a new solution in response to a unique customer query, that information can be added to the knowledgebase, which thus becomes "smarter."

SAP's CRM Software

A number of tools that provide CRM functionality exist within the SAP R/3 system. For example, to make sure that information about sales contacts is available throughout the organization, the SAP R/3 system provides a contact management tool, shown in Figure 3-13. This tool is essentially a database of personal contact information.

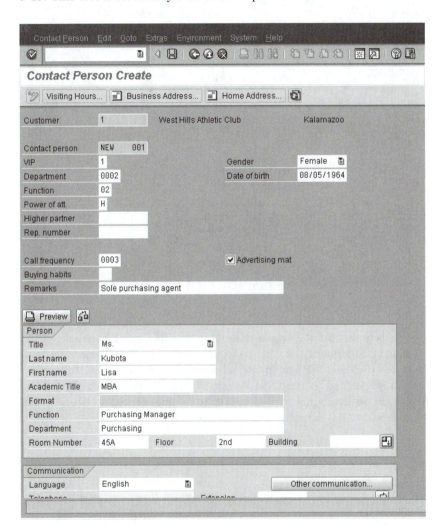

FIGURE 3-13   SAP R/3 contact manager

Another CRM tool in the SAP R/3 system is sales activity manager, an example of which is shown in Figure 3-14. This tool supports a strategic and organized approach to sales activity planning, and most importantly, can help make sure that follow-up activities are accomplished.

FIGURE 3-14 SAP R/3 sales activity manager

While the CRM tools in the SAP R/3 system, if employed properly, can help manage customer relationships, firms embracing the CRM concept often employ a separate CRM system that communicates with the ERP system. An advantage of this approach is that the planning and analysis performed in the CRM system do not interfere with the performance of the ERP system, which primarily processes large volumes of business transactions. SAP's mySAP provides additional customer interaction functionality.

Figure 3-15 shows how mySAP CRM relates to the SAP R/3 system as well as SAP's Business Warehouse (BW) and Advanced Planner and Optimizer (APO) systems.

FIGURE 3-15 SAP CRM system landscape

As described previously, the SAP R/3 system processes business transactions and provides much of the raw data for CRM. SAP's Business Warehouse is a flexible system for reporting and analysis of transactional data. By analyzing sales transactions using data mining, firms can discover trends and patterns to use in planning marketing activities. The Advanced Planner and Optimizer (APO) is a system that supports efficient planning of the supply chain. The APO's role in CRM is to provide higher levels of customer support through its Global Available-to-Promise (ATP) capabilities. In the standard SAP R/3 system, the ATP capabilities function on a location-by-location basis. If the product or material a customer wants is not available in the location that usually serves the customer, then the sales order clerk can check for the material in other facilities, but this must be done on a facility-by-facility basis. With Global ATP, the system automatically checks all facilities and determines the most cost-efficient facility to use to meet the customer's request. The mySAP CRM system communicates with the SAP R/3, BW, and APO systems in developing and executing its plans.

SAP's view of CRM is to provide a set of tools to manage the three basic task areas, or jobs, related to customers: Marketing, Sales, and Service. With CRM, these task areas contribute to the cultivation of the customer relationship. This cultivation goes through four phases as defined by SAP: Prospecting, Acquiring, Servicing, and Retaining.

In Prospecting, a potential new customer (or potential new business with existing customers) is evaluated and development activities (e-mails, sales calls, mailings, etc.) are planned to develop the prospective business. Marketing tasks predominate in this phase.

In Acquiring, salespeople develop business prospects into customers. Marketing is still the critical task, but the Sales tasks (processing inquiries, quotes, and eventually sales orders) become increasingly important in this phase.

Once sales are established with the customer, the business becomes one of servicing the account. Sales tasks are still important, but service tasks (including technical support, warranty work, product returns, quality problems, and complaint handling) are critical to maintain satisfied customers.

The rate at which a prospect becomes a customer is quite low, thus, a critical part of the process is Retention. It is much easier to retain a good customer than find a new one, so making sure that current customers are satisfied by timely delivery of quality products and services at a fair price is the focus of Retention. Sales and Service tasks are still critical, but Marketing tasks are again important to anticipate changes in the customer's requirements.

The customer development cycle (Prospecting, Acquiring, Servicing, and Retaining) is supported by Contact Channels, the methods the company uses to communicate with its customers. An interaction center provides contact through a variety of media (phone, fax, and e-mail). The Internet and mobile technologies are providing an increasingly large percentage of customer contacts. For example, the Contact Channels might aid a customer service agent on the telephone and prompt her to ask the customer various questions relating to the account.

Another set of tools in mySAP CRM is Marketing and Campaign Management. Companies invest significant sums of money in marketing campaigns, which are promotional activities that publicize the product and company. Successful planning, execution, and evaluation are critical to gain maximum effect from these efforts. Figure 3-16 shows how mySAP CRM supports marketing and campaign management. The top half of this diagram represents planning activities, while the bottom half represents execution and evaluation activities.

Marketing and Campaign Planning includes task scheduling, resource allocation, and budgeting. These planning tasks are executed in conjunction with Target Group Creation tasks, at which point data from the SAP R/3 system (perhaps using BW) are used to categorize the company's customers to offer more individual product and service promotions.

Campaign Execution Activity Management is a set of tools to help manage the execution of the marketing campaign, which can include handling sales calls, mailings, personalized e-mailings, and Web-based promotional activities directed to the group of customers at whom the promotion is targeted. These activities can be monitored throughout the campaign to make sure they are completed. For example, the status of a planned customer phone call will remain open until the call is completed.

FIGURE 3-16 Marketing and campaign planning

The Campaign Analysis tool allows the company to evaluate the success of the campaign so that it can incorporate improvements in the next marketing campaign. Marketing employees can use a number of measures to determine the success of the campaign, including tracking lead generation and response rates. Staff can use BW tools, by running queries, to support this analysis.

The Benefits of CRM

CRM provides companies with these benefits:

- *Lower costs:* CRM can lead to operational efficiencies, such as better response times in call center operations and better use of sales force time, which lower costs.
- *Higher revenue:* By segmenting customers, better selling opportunities present themselves and revenue increases.
- *Improved strategy and performance measurement:* Installing and operating an ERP system requires management and staff to think of the company as a whole. This attitude carries over into CRM work. With CRM in place, management can think about different performance measures; for example, should salespeople be rewarded for exceeding sales quotas and marketing people rewarded for finding new customers? Or, should both receive rewards that are based on some measure of customer satisfaction? The former approach—typical in days before CRM and ERP—can lead to unintegrated functional thinking. The latter approach—now feasible with CRM and ERP—can lead to all personnel thinking in terms of a company-wide effort to satisfy customers.

ANOTHER LOOK

CRM Success and Failure

Many organizations that have implemented CRM have not found the savings expected from the software. Gartner Inc. estimates that 65 percent of CRM implementations fail.[2] Part of the reason for these failures is that CRM is often incorrectly treated as another technology implementation driven by the IT department.[3] While there may be questions as to how failure is measured, these results have not diminished the interest in CRM. Companies are looking at new ways to reap the benefits from CRM, especially in the area of business intelligence. With a proper understanding, CRM is not considered simply a piece of software, but rather a business strategy. Companies that consider it only an IT implementation usually face failure—as with all IT projects, CRM must be aligned with the business strategy.[4]

CRM tools can be used to identify which customers are bringing in the most profits for the organization. It might be acceptable to lose a customer who brings in little revenue. The initial introduction of CRM software often identified new customers and helped organizations hold on to current ones. The newer releases of CRM use business intelligence tools that can help companies become more profitable.[5] For example, Volvo Cars of NorthAmerica is using predictive modeling to break down data on potential customers into clusters, compare the groups with current Volvo drivers, and hunt for new sales.[6] Predictive modeling is a data mining technique that gathers data, creates a statistical model on that data, makes predictions, and finally revises the model as new data flows in.

One model user of CRM technology is the European retailer Tesco. Tesco's success lies in the fact that it intensely analyzes the data, and then combines that analysis with exceptional customer service in a way that other retailers do not. The technical group at Tesco analyzes the data, and the business group interprets what that data analysis says about the customer. The collected data can influence the sorts of stores that should be developed, the locations of those stores, and the differing products offered from one geographic area to another.[7] Tesco's success is apparent: it is going beyond groceries and is also selling products such as vacations, cars, and insurance.

For financial services companies, CRM can help serve customers better and identify which are the best customers, ultimately becoming customer-centric. The financial services industry used to be a one-time seller of a product, such as a credit card. Now companies are looking for that customer relationship to sell further products. For example companies are interested in knowing what types of products customers are likely to be interested in buying and when they will want to buy them.

continued

In recognition that CRM programs are not easy to get right the first time, financial services software provider Dunstan Thomas has developed some guidelines for successful CRM:

1. CRM strategy definition: How do you see your company versus the competition? How do you define your customers?

2. Business process management: Examine your business processes and stream-line those using today's technology, be it the Internet or analysis tools such as data mining.

3. Back office integration: Your systems behind the scenes need to be robust and capable of handling all the requests. Databases should be "clean" and accessible.

4. Back-to-front office integration: The people answering questions need to get to the information instantly.

5. Realign the organization behind the customer: Employees need to recognize this customer focus and be rewarded for achieving this goal.[8]

Questions:

1. Assume you are the Chief Information Officer (CIO) of a large retail company that is thinking of implementing a CRM package, but there is resistance from the CEO. Do you think implementing CRM is a good idea? How can you convince the CEO that CRM will make improvements?

2. In what ways are companies now using CRM?

Chapter Summary

1. Fitter Snacker's unintegrated information systems are at the root of an inefficient and costly sales order process. Because information is not shared in real time, customers are asked to repeat initial sales order information. As an order is processed, errors in pricing, credit checks, and invoicing also occur, presenting a poor company image to customers. Integrated ERP software would let FS avoid errors because all customer data are stored in a central database that is shared in real time by all company employees.

2. An ERP system such as SAP R/3 sees a sale as a sequence of related functions, including taking orders, setting prices, checking product availability, checking the customer's credit line, arranging for delivery, billing the customer, and collecting payment. In SAP R/3 all these transactions, or documents, are electronically linked, so tracking an order's status (partial shipments, returns, partial payments, and so forth) is easily accomplished.

3. When an ERP system is installed, various configuration decisions are made. These decisions reflect management's view of how transactions should be recorded and later used for decision making. For example, the system can be configured to limit selling price discounts, thus avoiding unprofitable pricing.

4. An ERP system's central database has master data tables, which store relatively permanent data about customers, suppliers, material, and inventory as well as transaction data tables, which store relatively temporary data such as sales orders and invoices.

5. Customer relationship management (CRM) systems build on the organizational value ERP provides; they specifically increase the flexibility of the company's common database in regards to customer service. Various kinds of CRM software are available, some from ERP vendors (including SAP) and some from third-party software companies. CRM software can lead to operational savings, but most companies buy it because they feel that having better customer relationships will result in higher revenues. Uses of CRM have evolved since the software was initially launched, beginning as a customer contact repository to extending its capabilities to sophisticated business intelligence.

Key Terms

Audit trail	Delivery
Condition technique	Document flow
Customer master data	Material master data
Customer relationship management (CRM) software	Organizational structure

Exercises

1. Assume you are the Fitter Snacker saleperson calling on your local grocery store. You just started this job, and you are nervous about meeting your customer for the first time. Describe all the problems you might encounter, using the old Fitter Snacker process described in this chapter, when taking and filling the order. Now assume you have an ERP system in place. Describe the sequence of events in selling the snack bars to your local grocery store with the new system in place.

2. Assume Fitter Snacker plans to install an ERP system and reorganize its sales divisions, consolidating them into one. What personnel issues would the management have to deal with when promoting the new system to the sales force?

3. Fitter Snacker's current sales order accounting involves recording sales in each sales division and then periodically sending enough data to Accounting to record sales for the company. Complete sales order data are retained in each sales division for business analysis purposes. Assume that different divisions of the Delicious Foods Company buy NRG-A and NRG-B bars from each of FS's sales divisions. To complicate matters, some divisions of Delicious buy store-brand bars from FS. (Delicious owns convenience store outlets.) FS management wants to see an analysis of the overall relationship with Delicious Foods. FS's management thinks there may be opportunities to promote the relationship with Delicious, but they need to assess profitability before proceeding. They want to see what products each division sells to Delicious, how much is sold, and the terms. Assume in FS's current system that all the required data are available only at the sales division level. What steps will be needed to pull this company-wide analysis together? (Review how each division sells its products and keeps its records.) Do you think a sales division manager will be enthusiastic about sharing all data with his or her counterpart in the other division? Or, do you think there might be some reluctance? Why?

4. Continuing the Delicious Foods example, now assume that FS has a SAP R/3 system installed. Each sales division records sales in the same way. Sales records exist in real time and are kept in the company's common database. What steps will be needed to pull this company-wide analysis together? Do you expect that the divisions will meet the new system with enthusiasm or reluctance?

5. Assume you are the CIO for Fitter Snacker. You need to convince the top management to approve an ERP implementation project. Write a persuasive memo to top management, convincing them of this need. Include specific examples of ERP benefits in the memo, such as the elimination of the credit check time delay.

6. Describe how FS's SAP system simplifies looking up customer numbers, setting a delivery date, and charging a unique price to a given customer.

7. What are master data and why are master data an important part of ERP? What is document flow and why is it important?

8. Do ERP systems strengthen or weaken a corporate manager's control over operations? State your point of view and justify it with examples from this chapter.

9. How can a business better serve its customers using the APO tool in SAP?

10. Assume you are the marketing manager for a large consumer products company. You need to launch a new marketing campaign. How can you convince your company of the value of using CRM to help with this campaign?

11. Investigate uses of data mining on the Internet, and report on companies' experience with it.

12. Research articles about information systems on the Web to find out what companies are doing with data warehouses.

End Notes

[1] Chris Murphy, "IT Confidential Executive Swaps and Expletives," *Information Week* (August 9, 2004).

[2] "The CRM Journey," http://www.databasesystemscorp.com/tech-crm_26.htm.

[3] "The CRM Journey."

[4] "CRM Must Be Linked to Value: An Interview with Francis Buttle," CRMGuru.com (June 17, 2004).

[5] "Business Intelligence Can Give CRM A Boost," *Information Week* (June 18, 2004).

[6] "Head Games," (Special Report CRM) *CFO* (July 2004).

[7] "CRM Must Be Linked to Value: An Interview with Francis Buttle."

[8] "The CRM Journey."

Supplemental Resources

The following articles were also used as resources:

Don Peppers and Martha Rogers, "Customer Value," *CIO* (September 15, 1998) (archived at *http://www.cio.com/archive/enterprise/091598_hs.html*).

Louise Fickel, "Know Your Customer," *CIO* (August 15, 1999) (archived at *http://www.cio.com/archive/081599_customer.html*).

Carol Hildebrand, "One to a Customer," *CIO* (October 15, 1999) (archived *at http://www.cio.com/archive/enterprise/101599_customer.html*).

Lee Pender, "CRM from Scratch," *CIO* (August 15, 2000) (archived at *http://www.cio.com/archive/081500_scratch.html*).

Elana Varon, "Suite Returns," *CIO* (August 15, 2000) (archived at *http://www.cio.com/archive/08152000_suite.html*).

Mindy Blodgett, "Masters of the Customer Connection," *CIO* (August 15, 2000) (archived at *http://www.cio.com/archive/081500_overview.html*).

John Edwards, "What's Your Problem?" *CIO* (September 1, 2000) (archived at *http://www.cio.com/archive/090100_problem.html*).

Stewart Deck, "Crunch Time," *CIO* (September 15, 2000) (archived at *http://www.cio.com/archive/09152000_crunch.html*).

CHAPTER 4

PRODUCTION AND SUPPLY CHAIN MANAGEMENT INFORMATION SYSTEMS

LEARNING OBJECTIVES

After completing this chapter, you should be able to:

- Describe the steps in the production planning process of a high-volume manufacturer such as Fitter Snacker.
- Describe Fitter Snacker's production and materials management problems.
- Describe how a structured process for supply chain management planning enhances efficiency and decision making.
- Describe how production planning data in an ERP system can be shared with suppliers to increase supply chain efficiency.

INTRODUCTION

In Chapter 2, you learned that Enterprise Resource Planning (ERP) has its roots in Materials Requirements Planning (MRP). In fact, MRP is still a large part of today's ERP systems. In this chapter, we'll look at supply chain management (SCM) in an ERP system. Fitter Snacker is part of a supply chain that starts with farmers growing oats and wheat and ends with a customer buying an NRG bar from a retail store. First, we will examine how Fitter Snacker manages its production activities, and then we will look at the broader concept of the supply chain management (SCM).

In Chapter 3, we looked at Fitter Snacker's sales order process, and we assumed that FS had enough warehoused snack bars to fill an order. Like most unintegrated manufacturing operations, however, FS often has trouble scheduling production. Consequently, sometimes its warehouse is not properly stocked. In this chapter, you'll explore FS's SCM problems, and you'll see how ERP can help fix them.

PRODUCTION OVERVIEW

To meet customer demand efficiently, Fitter Snacker must develop a forecast of customer demand, then develop a production schedule to meet the estimated demand. While developing a production plan is not a trivial task, the end result answers two questions:

1. How many of each type of snack bar should we produce, and when?
2. What quantities of raw materials should we order so we can meet that level of production, and when should they be ordered?

But a good production plan is not enough: Fitter Snacker must be able to execute the plan and make adjustments when customer demand inevitably does not meet the forecast. An ERP system is a good tool for developing and executing production plans because it integrates the SCM functions of Production Planning, Purchasing, Materials Management/ Warehousing, Quality Management, and Sales and Accounting. In addition, ERP systems can be connected to supplier and customer information systems to support even better planning of the supply chain.

In this chapter, we will use spreadsheet examples to illustrate the logic that FS should be using to plan and schedule production of the bars. First, the production process of manufacturing the bars is explained. After demonstrating the planning and scheduling logic at each stage of the production planning process using spreadsheets, we will show the SAP R/3 screens that implement the logic in an ERP environment. Throughout the chapter, you will learn why using an integrated information system is superior to using an unintegrated system.

The goal of production planning is to schedule production economically so that goods can be shipped to customers to meet promised delivery dates in the most cost-efficient manner. There are three general approaches to production:

1. *Make-to-stock* items are made for inventory (the "stock") in anticipation of sales orders. Most consumer products (for example, cameras, canned corn, and books) are made this way.
2. *Make-to-order* items are produced to fill specific customer orders. This approach is usually taken when producing items that are too expensive to keep in stock or items that are made or configured to customer specifications. Examples of make-to-order items are airplanes and large industrial equipment.
3. *Assemble-to-order* items are produced using a combination of make-to-stock and make-to-order processes. The final product is assembled for a specific

order from a selection of make-to-stock components. Personal computers are a typical assemble-to-order product.

Fitter Snacker's Manufacturing Process

Fitter Snacker uses make-to-stock production to produce its snack bars. The manufacturing process is illustrated in Figure 4-1.

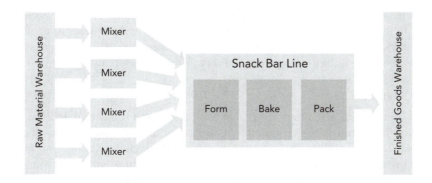

FIGURE 4-1   Fitter Snacker's manufacturing process

The snack bar line can produce 200 bars a minute, or 3,000 pounds of bars per hour. Each bar weighs 4 ounces. The entire production line operates on one shift a day. Here's how it works.

Fitter Snacker's Production Sequence

Raw materials are taken from the warehouse to one of four mixers. Each mixer mixes dough in 500-pound batches. Mixing a batch of dough requires 15 minutes of mixing time, plus 15 additional minutes to unload, clean, and load the mixer for the next batch of dough; therefore, each mixer can produce two 500-pound batches of dough per hour—more than the production line can process. Because only three mixers need to be operating at a time to produce 3,000 pounds of snack bars per hour, a mixer breakdown will not shut down the production line.

After mixing, the dough is dumped into a hopper at the beginning of the snack bar line. A forming mechanism molds the dough into bars, which will weigh 4 ounces each after baking. Next, an automated process takes the formed bars on a conveyor belt through an oven that bakes the bars for 30 minutes. When the bars emerge from the oven, they are individually packaged in a foil wrapper. A group of 24 bars are packaged into a display box. At the end of the snack bar line, production personnel either stack the display boxes on pallets (for small orders) or pack the display boxes into shipping boxes and take them to the finished goods warehouse.

Changing the snack bar line from one type of bar to the other takes 30 minutes for cleaning the equipment and changing the wrappers, display boxes, and shipping cases. Each night, a second shift of employees cleans all the equipment thoroughly and sets it up for the next day's production. Thus, changing production from NRG-A to NRG-B bars can be done at the end of each day without a loss of capacity. On the other hand, if two products are produced in one day, then a half-hour of capacity is lost in the changeover.

Fitter Snacker's Production Problems

Fitter Snacker has no problems *making* snack bars. Fitter Snacker has problems deciding *how many* bars to make and *when* to make them. The manufacturing process at FS suffers from a number of problems, from communication and inventory to accounting issues, mainly stemming from the unintegrated nature of its information systems.

Communication Problems

Communication breakdowns are inherent in most companies, and they are magnified in a company with an unintegrated IS. For example, FS's Marketing personnel do not share information with Production personnel: Marketing frequently excludes Production from meetings, doesn't consult Production when planning sales promotions, and fails to notify Production of planned promotions. Similarly, Sales does not warn Production that it has taken an exceptionally large order.

When Production must meet an unexpected increase in demand, several things happen. First, an unexpected spike in sales depletes warehouse inventories. To compensate, Production must schedule overtime production, paying for overtime labor, which results in higher production costs for products. In addition, some materials, such as ingredients, wrappers, and display boxes, are custom products purchased from a single vendor. A sudden increase in sales demand might cause shortages or even a stockout of these materials. Getting these materials to the plant might require expedited shipping, further increasing the cost of production. Finally, Production personnel are evaluated on their performance, such as how successful they are at controlling costs, keeping manufacturing lines running, quality control, and safety. If they can't keep production costs down, they receive poor evaluations. Managers are especially frustrated when an instant need for overtime follows a period of low demand, when inventory could have been built up in anticipation of an increase in sales following a product promotion by marketing.

Inventory Problems

Week-to-week and day-to-day production planning is not linked to expected sales levels in a systematic way. When deciding how much to produce, the production manager applies rules developed through experience. Her primary indicator is based on the normal amount of finished goods inventory that should be stocked and the actual inventory levels of finished goods in the warehouse. Thus, if NRG-A or NRG-B inventory levels seem low, the production manager schedules more bars for production. Her judgment is also influenced by what information she might hear informally from people in Marketing about expected sales.

The production manager's inventory data are in an Access database. Data records are not maintained in real time, and they do not flag inventory that has been sold but not yet shipped. Such inventory is not available for sale of course, but this discrepancy cannot be determined from the records. Thus, the level of inventory that is available to ship is not known. This is problematic if the Wholesale Division generates unusually large orders or a high volume of orders. For example, two large Wholesale Division orders arriving at the same time can deplete the entire available inventory of NRG-A bars. In this case, the Production department must change the production schedule for NRG-B bars so it can fill the orders for NRG-A. This changeover means production of NRG-B bars must be delayed, and production capacity will be lost due to the unplanned production changeover.

The production manager lacks not only a systematic method for meeting anticipated sales demand, but also a systematic method for adjusting production to reflect actual sales. Data on actual sales are not shared with the Production department, partly because it is hard to gather on a timely basis and partly because of a lack of trust between the Sales and Production departments (this distrust may exist as a result of prior negative experiences or competition between the departments). Hence, the production manager must use warehouse inventory levels as a benchmark; standard levels of inventory are the only guide available. If Production had access to sales forecasts and real-time sales order information, then timely adjustments to production could be made if needed. These adjustments would allow for inventory levels to be much closer to what is actually needed.

Accounting and Purchasing Problems

Production and Accounting do not have a good way to accumulate the day-to-day costs of production. As discussed in Chapter 3, the warehouse keeps a fairly good running count of what is on hand. Furthermore, there is a monthly inventory taken, and actual counts are usually within 5 percent of what is on the books. Management would like to be more accurate, of course, but without a real-time inventory system, that goal cannot be met.

Manufacturing costs are based on the number of bars produced each day, a number that is measured at the end of the snack bar production line. Fitter Snacker uses standard costs for the purpose of figuring manufacturing costs. **Standard costs** are the cost it normally takes to manufacture a product and are calculated from historical data and any changes in manufacturing that have occurred since the historical data were collected. For each batch of bars it produces, FS can estimate materials cost, labor cost, and related factory overhead. The number of batches is counted and multiplied by the standard cost of a batch. The resulting amount is charged to manufacturing costs.

Most manufacturing companies use standard costs in some way. The method, however, requires that standards be adjusted periodically to conform with actual costs. Actual Fitter Snacker raw material and labor costs, however, often deviate from the standard costs. FS is not good at controlling raw materials purchases, and the production manager cannot give the purchasing manager a good production forecast. So the purchasing manager works on two tracks: first, she tries to keep inventories high to avoid stockouts. Second, if she's offered good bulk-quantity discounts on raw materials like oats, she will buy in bulk, especially for items that have long lead-times for delivery. These purchasing practices make it difficult to forecast the volume of raw materials that will be on hand and their average cost. Similarly, the frequent need for overtime labor means that FS has trouble accurately forecasting the average cost of labor for a batch of bars.

Thus, Production and Accounting must occasionally compare standard costs with actual costs and then adjust the accounts for the inevitable differences. This is always a tedious and unpleasant job. The comparison should be done at each monthly closing, but often it is put off until the closing at the end of each quarter, when FS's financial backers require legitimate financial statements. The adjustments are often quite large, depending on production volumes and costs during the quarter.

Exercise 4.1

a. A convenience store chain offers to buy a very large amount of its "store" brand health bars, which is the NRG-B bar with a different wrapper. The chain wants a lower-than-normal selling price, but it is also offering to place a very large order, one that is quadruple the size of a regular order. The marketing manager asks managers from Production, Purchasing, and Accounting whether the terms of the proposed deal will be profitable. Why will the managers in these areas have trouble providing a reliable answer on short notice?

b. The production manager notes that warehouse inventory levels are fairly high, so the production line does not need to be run for a full eight hours each day during the coming week. Workers would still need to be paid during the idle time, however, and overhead costs would be incurred as well. The production manager plans to run the line for eight hours a day anyway, because this will decrease the average cost of bars actually produced. (Costs would be spread over more bars.) Furthermore, some warehoused raw materials will spoil if they're not used soon. Is the production manager's reasoning logical? Why or why not?

THE PRODUCTION PLANNING PROCESS

In this section, you will examine a systematic process for developing a production plan that takes advantage of an ERP system. Production planning is a complicated process. To explain this process, spreadsheet calculations are presented to illustrate the key steps in developing a production plan, followed by the corresponding screens in the SAP R/3 system.

Production planners follow three important principles:

1. Work from a sales forecast and current inventory levels to create an "aggregate" production plan for all products. Aggregate production plans help to simplify the planning process. The plans are usually aggregated in two ways. First, plans are made for groups of related production rather than for individual products. Second, the time increment used in planning is frequently a month or a quarter rather than the daily or weekly production plans that will actually be executed. Aggregate plans should consider the available capacity in the facility.

2. Break down the aggregate plan into more specific production plans for individual products and smaller time intervals.

3. Use the production plan to determine raw material requirements.

Production planners aggregate products into product groups to reduce the number of variables they must consider when developing a production plan. Developing production groups can become complicated. For example, a cereal manufacturer can group together different package sizes, or they might group together product brands (such as kid cereals, health cereals, and so on) or product types (in the case of a consumer products company: shampoo, laundry detergent, and disposable diapers). For Fitter Snacker, the only two products, NRG-A and NRG-B bars, will be combined into one group to illustrate the process. In addition, the aggregate production plan will be developed using a monthly time increment. This monthly production plan will be disaggregated to determine weekly raw material orders and daily production schedules.

The SAP R/3 Approach to Production Planning

SAP R/3's approach to the production planning process is shown in Figure 4-2. Refer to this figure throughout this section to trace production planning.

FIGURE 4-2   The production planning process

The information at each stage of the production process flows through the following steps. Each step is explained in detail in the next section of the chapter.

- *Sales Forecasting* is the process of developing a prediction of future demand for a company's products.
- *Sales and Operations Planning (SOP)* is the process of determining what the company will produce. The Sales Forecast and Starting Inventory levels are inputs to this process. At first glance, it would seem that a company should just make products to match the forecasted sales, but developing the production

plan can be complicated because capacity must be considered. Many products have seasonal demand, and to meet demand during peak periods, production planners must decide whether to build up inventory levels before the peak demand, increase capacity during the peak period, subcontract production, or use some combination of these approaches.

80

- *Demand Management* is the process of breaking down the production plan into finer time units, such as weekly or even daily production figures, to meet demand for individual products.

- *Detailed Scheduling* uses Demand Management's production plans as an input for a production schedule. Methods of detailed scheduling depend on the manufacturing environment. For Fitter Snacker, the detailed production schedule will determine when to switch between the production of NRG-A and NRG-B bars.

- *Production* uses the detailed schedule to manage daily operations, answering the questions, "What should we be producing?" and "What staffing do I need?"

- *Materials Requirements Planning (MRP)* determines the amount and timing of raw material orders. This process answers the questions, "What raw materials should we be ordering so we can meet a particular level of production, and when should we order them?"

- *Purchasing* takes the quantity and timing information from MRP and creates purchase orders for raw materials, which it transmits to qualified suppliers.

Let's take a more detailed look at each of these steps in the production process.

Sales Forecasting

A range of forecasting techniques can be used to predict consumer demand. Fitter Snacker has no formal way of developing a sales forecast and sharing it with Production. SAP's R/3 system takes an integrated approach. Whenever a sale is recorded in the Sales and Distribution (SD) module, the quantity sold is recorded as a consumption value for that material. These consumption values can be maintained on a weekly or monthly basis, as desired. If more detail is needed, the Logistics Information System that is part of SAP can record sales with more detail (for example, by region or sales office) or data can be stored in the separate Business Warehouse (BW) system for more detailed analysis. With an integrated information system, accurate historical sales data are available for forecasting.

One simple forecasting technique is to use a prior period's sales and then adjust those figures for current conditions. To make a forecast for Fitter Snacker, we can use the previous year's sales data in combination with marketing initiatives to increase sales. Look at the forecasts shown in Figure 4-3.

The sales data in Figure 4-3 are for shipping cases, which contain 12 display boxes that each contain 24 bars for a total of 288 bars. Note in Figure 4-3 that the forecast starts with the previous year's sales levels to reflect FS's seasonal sales fluctuations (sales are higher in the summer when more people are active). Also note that there was a special marketing promotion last year. The estimated impact of this promotion was an increase in sales of 300 cases for May and June. This increase must be subtracted from the previous year's sales values to get an accurate base measurement. FS's Marketing department anticipates a 3 percent growth in sales over the previous year, based on research reported in trade

Sales Forecasting	Jan.	Feb.	March	April	May	June
Previous Year (cases)	5734	5823	5884	6134	6587	6735
Promotion Sales (cases)					300	300
Previous Year base (cases)	5734	5823	5884	6134	6287	6435
Growth: 3.0%	172	175	177	184	189	193
Base Projection (cases)	5906	5998	6061	6318	6476	6628
Promotion (cases)						500
Sales Forecast (cases)	5906	5998	6061	6318	6476	7128

FIGURE 4-3   Fitter Snacker's sales forecast for January through June

publications. And finally, FS will be launching a special marketing promotion at the end of May to increase sales at the beginning of the summer season. FS marketing experts think this will result in an increase in sales of 500 cases for June.

Exercise 4.2

Following the format of the spreadsheet shown in Figure 4-3, develop a spreadsheet to forecast Fitter Snacker's sales for July through December. Make the sales growth rate of 3 percent an input value, and calculate the base projection using the previous year's values shown in Figure 4-4. Assume that the special marketing promotion last year resulted in an increase in sales of 200 cases for July, and a special marketing promotion this year will result in an increase in sales for July of 400 cases.

Sales Volume	July	Aug.	Sept.	Oct.	Nov.	Dec.
Previous Year	6702	6327	6215	6007	5954	5813

FIGURE 4-4   Fitter Snacker's sales for the previous period, July through December

Sales and Operations Planning

Sales and Operations Planning (SOP) is the next step in the production planning process. The input to this step is the sales forecast provided by Marketing. The output of this step is a production plan designed to balance market demand with production capacity. This plan is the input to the next step, Demand Management.

Developing a sales and operations plan from a sales forecast answers the question: "How can Manufacturing efficiently produce enough goods to meet projected sales?" In Fitter Snacker's case, there is no way to answer this question, because FS does not produce a formal estimate of sales. If FS had an ERP system, the calculation would be done as described here.

We know that Fitter Snacker can produce 200 bars per minute, so we can estimate the production capacity required by the sales forecast. Figure 4-5 shows FS's Sales and Operations Plan for the first six months of the year.

At the start of January, the production planner is projecting a beginning inventory of 100 cases. The first line in Figure 4-5 is the Sales Forecast, which is the output of the Sales forecasting process shown in Figure 4-3. The next line is the Production Plan, which is developed by the production planner in a trial-and-error fashion, observing the effect of different production quantities on inventory levels and capacity utilization. The goal is to

Sales and Operations Planning		Dec.	Jan.	Feb.	March	April	May	June
1) Sales Forecast			5906	5998	6061	6318	6476	7128
2) Production Plan			5906	5998	6061	6318	6900	6700
3) Inventory		100	100	100	100	100	524	96
4) Working Days	.		22	20	22	21	23	21
5) Capacity (Shipping Cases)			7333	6667	7333	7000	7667	7000
6) Utilization			81%	90%	83%	90%	90%	96%
7) NRG-A (cases)	70.0%		4134	4199	4243	4423	4830	4690
8) NRG-B (cases)	30.0%		1772	1799	1818	1895	2070	2010

FIGURE 4-5 Fitter Snacker's sales and operations plan for January through June

develop a production plan that meets demand without exceeding capacity and that main-tains "reasonable" inventory levels (neither too high or two low). This process requires judgment and experience. The third line, Inventory, is the difference between the Sales Forecast and the Production Plan. The production planner has developed a plan that maintains a minimum planned inventory of 100 cases. This inventory, called safety stock, is planned so that if sales demand exceeds the forecast, sales can be met without alter-ing the production plan. Notice that in May, the Production Plan is greater than the May Sales Forecast, and the inventory is 524. Why? Because the planner wants to build up inventory to handle the increased demand in June, which results from the normal sea-sonal increase in snack bar sales and additional demand from the planned promotional activities. The fourth line is Working Days, an input based on the company calendar. Using the number of working days in a month, the available capacity each month is calcu-lated in terms of the number of shipping cases.

> 200 bars per minute × 60 minutes per hour × 8 hours per day = 96,000 bars per day, which equals 333.3 cases per day (96,000 bars per day ÷ 24 bars per box ÷ 12 boxes per case).

If you multiply the number of working days in a month by the production capacity of 333.3 shipping cases per day, you get the monthly capacity in shipping cases, which is shown in line 5.

With the available Capacity (assuming no overtime) now expressed in terms of ship-ping cases, it is possible to determine the capacity utilization for each month by dividing the Production Plan (line 2) by the available Capacity (line 5) and expressing the result as a Utilization percentage (line 6). This capacity calculation lets us see whether we have the capacity necessary to meet the production plan. While higher levels of capacity utiliza-tion mean that Fitter Snacker is producing more with its production resources, this per-centage must be kept below 100 percent to allow for production losses due to product changeovers, equipment breakdowns, and other unexpected production problems.

The last step in Sales and Operations Planning is to disaggregate the plan, that is, break it down into plans for individual products. Lines 7 and 8 disaggregate the planned produc-tion shown in line 2, based on the breakdown of 70 percent NRG-A and 30 percent NRG-B snack bars. This 70/30 breakdown is based on previous sales data for these products. These monthly production quantities are the output of the Sales and Operations Plan-ning process, and are the primary input to the Demand Management process.

Suppose that FS is regularly able to achieve production levels at 90 percent of capacity. If the sales forecast requires more than 90 percent capacity, FS management can choose from among the following alternatives to develop a production plan.

1. FS might choose not to meet all the forecasted sales demand, or it might reduce promotional activities to decrease sales.
2. To increase capacity, FS might plan to use overtime production. Doing that, however, would increase labor cost per unit.
3. Inventory levels could be built up in earlier months, when sales levels are lower, to reduce the capacity requirements in later months. Doing that, however, would increase inventory holding costs and increase the risk that NRG bars held in inventory might pass their expiration date before being sold by retailers.
4. Management might try a hybrid approach to the capacity problem to find the right balance: reduce sales promotions slightly, increase production in earlier months, and plan for some overtime production.

The monthly production quantities in lines 7 and 8 of Figure 4-5 creates some inventory in May to meet June's sales; in addition, some overtime production is likely in June because capacity utilization is over 90 percent. This illustrates the value of an integrated system: it provides a tool to incorporate data from Marketing and Manufacturing to evaluate different plans. Whereas Marketing may want to increase sales, the company might not increase its profits if overtime costs or inventory holding costs are too high. This sort of planning is difficult to do without an integrated information system, even for small companies like FS. Having an integrated information system helps managers keep corporate profit goals in mind.

In SAP R/3, the sales forecast can be made using historical sales data from the Sales and Distribution (SD) module or it can be generated from plans developed in the Controlling (CO) module. In the CO module, profit goals for the company can be set, which can then be used to estimate the sales levels needed to meet the profit goals. Figure 4-6 shows the Sales and Operations Planning screen from the SAP R/3 system. The title of this screen is "Create Rough-Cut Plan." **Rough-cut planning** is a common term in manufacturing for aggregate planning. As described above, rough-cut plans are disaggregated to generate detailed production schedules.

The sales forecast is entered in the first row (Sales) of the rough-cut plan. Data can be entered manually by the user, a sales forecast can be transferred from a profitability analysis performed in the CO module, or the user can perform a forecast in this screen, calling up historical sales data from the SAP R/3 system. The second row, Production, represents the production that is planned to meet the sales forecast. It, too, can be entered manually, or the SAP system can generate values that meet sales goals. The third row shows inventory as the Stock level. The gray color indicates that it is a calculated result. The fourth row allows for the entry of a Target stock level. If the user enters a value in this row, then SAP R/3 will propose production levels to meet the Target stock level. The sixth row lets the user specify a target stock level in terms of the number of days of demand it would cover, known as Target Days Supply. The SAP system uses the Factory Calendar that specifies company holidays and planned shutdowns to determine the number of working days in a month when calculating the Target Days Supply.

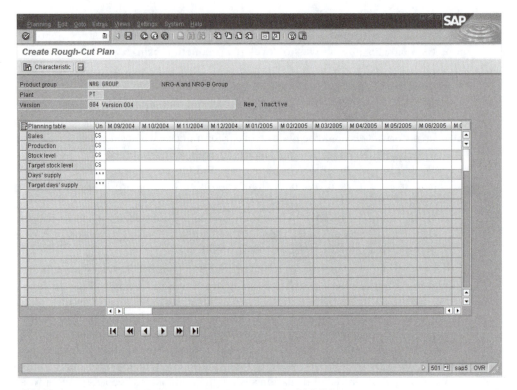

FIGURE 4-6   Sales and Operations Planning screen in SAP R/3

If the sales plan (first line in Figure 4-6) is to be developed using forecasting tools, the SAP R/3 system will provide the planner with historical sales values based on sales data stored in the system. Without an integrated system like SAP, the planner would likely have to request sales figures from the sales department, and may not be sure how accurate the data were. Figure 4-7 shows how the SAP R/3 system displays historical sales figures. In addition to providing the historical sales values from the SD module, this screen allows the planner to "correct" the sales values. For example, sales may have been low in the past due to unusual weather conditions, or sales would have been higher if the company could have met all demand. The sales figures used for forecasting should represent the best estimate of what demand was in the past, not necessarily what the actual sales were.

The SAP R/3 system can automatically graph these data to help the planner determine if there are any unusual patterns in the historical sales values that might require investigation. The planner can "correct" these values as well. Adjustments might be made to sales values that were unusually high or low. The data can also be adjusted to "back out" the effects of previous sales promotions. After the forecast is made, the sales forecast can be adjusted to incorporate increased sales from planned sales promotions. Once the historical data are acceptable, the user can select one of the SAP R/3 tools shown in Figure 4-8 to prepare the forecast.

FIGURE 4-7   Historical sales levels for Fitter Snacker

FIGURE 4-8   Forecasting model options in SAP R/3

This screen allows the user to specify a number of forecasting parameters, including whether the model should allow for trends and seasonal variations. Once the SAP R/3 system generates a forecast, the planner can view the results graphically, as shown in Figure 4-9. While the SAP R/3 system also provides the standard statistical measures of forecast accuracy, human judgment is frequently the best determinant of whether the forecast results make sense.

FIGURE 4-9 Forecasting results presented graphically in SAP R/3

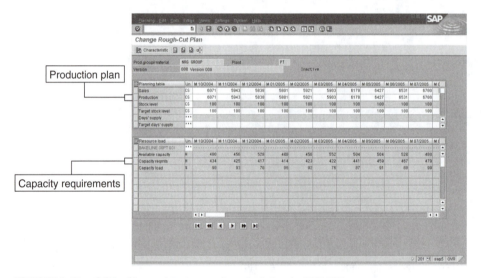

FIGURE 4-10 SOP with rough-cut capacity calculation in SAP R/3

Another feature of the SAP R/3 sales and operations planning process is the integration of rough-cut capacity planning, which is shown in Figure 4-10. Rough cut capacity planning applies simple capacity estimating techniques (like those shown in the spreadsheet example in Figure 4-5) to the production plan to see if they are feasible. Frequently, rough-cut capacity planning techniques are applied to critical resources—those machines or production lines where capacity is usually limited. For a company like Fitter Snacker with a simple manufacturing process, these estimates can be very accurate. For more complex manufacturing processes, these estimates will not be completely accurate but will ensure that the production plans are at least reasonable. Managers can use SAP R/3's more sophisticated planning tools at the detailed scheduling level, when the plans that are developed will actually be converted into manufacturing decisions on the shop floor.

While an integrated ERP system like SAP can provide sophisticated tools to support sales and operations planning, the plan will only be successful if the interested parties are committed to the process. If Marketing and Manufacturing cannot agree on sales forecasts, promotions, and production plans, then the company will find that it is overstocked in some items, running out of others, and spending a lot of money on overtime production and expedited shipping. Successful sales and operations planning depends on developing a culture of cooperation between Marketing and Manufacturing, which usually requires continuous support, encouragement, and enforcement from top management. As you will learn in Chapter 7, changing a company's culture is usually a much harder challenge than the technical challenge of installing new hardware and software.

ANOTHER LOOK

Sales and Operations Planning

The Kellogg Company has realized tremendous savings and efficiencies from a coordinated Sales and Operations Planning Process.

A key to the success of this effort was changing the focus of the key players. Previous to the new process, Marketing and Sales personnel were evaluated on how many tons of cereal they sold. Manufacturing was evaluated on how many tons of cereal they produced. Nobody was focused on whether they were making a profit. The Marketing area was focusing on national sales promotions, which created huge peaks in demand. Sales personnel were making sales deals that moved a lot of cereal, but at terms that didn't necessarily make profits for the company. Sales forecasts were often incorrect and last-minute customer orders created inefficient distribution and logistics practices as a product was often redeployed to fulfill these rush orders. Production personnel focused on making as much cereal as possible, which meant that they minimized product changeovers so they didn't lose production throughput and line capacity during the changeovers. These long production runs meant that a lot of cereal was being produced, but it wasn't necessarily the cereal that was in demand by the customer. With a focus on minimizing waste in the plants, cereal was often produced to avoid scrapping packaging, even though the demand forecast did not require the additional production. This also meant that overproduced and misdeployed production resulted in old age and damaged product that needed to be liquidated or written off as unsaleable goods.

The Kellogg Company now has a structured sales and operations planning process called "Integrated Business Planning" (IBP) with all participants focused on making profitable decisions for the company. Marketing has moved from national product promotions to customer-specific product promotions that are time-phased to reduce the up-and-down demand cycle and inventory peaks. Because the plans are coordinated with Manufacturing, the production schedules are developed to support the promotions and to produce to the forecast with a minimum of safety stock. As Manufacturing is now evaluated on whether the company makes money, it plans production so that it produces the

continued

products that are in the demand forecast and can be sold. With a profit goal, the sales force is motivated to negotiate deals that make money for the company and not deals at any cost merely to sell cereal. There is also an overall focus on value (dollar revenue) versus volume (production tonnage).

Developing an effective Sales and Operations Planning process at the Kellogg Company was not easy, but the rewards have been significant. The process is an important factor in the recent success of the company. In October 2004, the Kellogg Company Web site reported third quarter earnings growth driven by increased sales, investment, and effective execution. The company raised its earnings guidance for the full year of 2004 due to these strong results and continued business momentum. Because Marketing, Sales, and Manufacturing are working together along with Finance, the Kellogg Company has been able to reduce its production capacity, finished goods inventory, and capital requirements, while selling more cereal than ever before, because it is producing the right product in the right quantities at the right time.

The results have manifested themselves in fresher products and improved order fulfillment for the consumer, fewer stockouts for retailers, reduced damages and unsaleable merchandise, improved freight and warehouse costs, more efficient customer promotion investments, and significantly increased cash flow. These improvements combine to increase revenues, profits, share of market, and shareholder equity.

Question:

1. Write a memo to the CIO, manufacturing manager, and marketing manager of Fitter Snacker, convincing them of the importance of a sales and operations planning process. Use the Kellogg Company example to be persuasive.

Disaggregating the Sales and Operations Plan

As mentioned previously, companies typically develop sales and operations plans for product groups. Product groups are especially important for companies that have hundreds of products, because developing individual plans for hundreds of individual products is extremely time consuming. Furthermore, it would be hard to develop these plans in a coordinated fashion so that production capacity can be considered. Fitter Snacker's product group is very simple, consisting of 70 percent NRG-A and 30 percent NRG-B bars. Figure 4-11 shows how product groups are defined in the SAP R/3 system. The system allows any number of products to be assigned to a product group. In addition, product groups can have other product groups as members, so complicated aggregations can be defined.

When the Sales and Operations Plan (SOP) is disaggregated, the production plan quantities specified for the group are transferred to the individual products that make up the group according to the percentages defined in the product group structure (Figure 4-11).

The results of the disaggregation process can be seen in the Stock/Requirements List shown in Figure 4-12. This screen displays the inventory level for an individual product, including all planned additions and reductions. As you can see in Figure 4-12, the production plan from Sales and Operations Planning has been added to the Stock/Requirements List for NRG-A bars as reductions to the inventory levels.

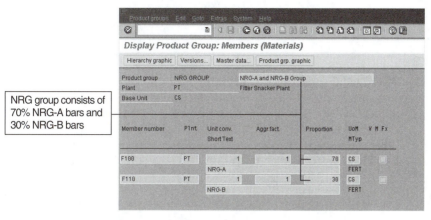

NRG group consists of 70% NRG-A bars and 30% NRG-B bars

FIGURE 4-11   Product group structure in SAP R/3

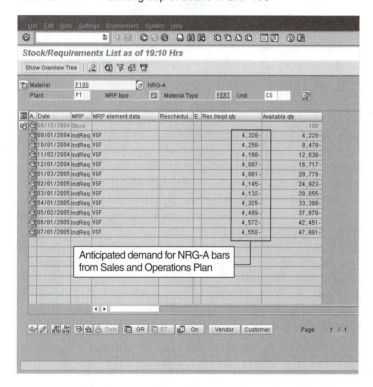

Anticipated demand for NRG-A bars from Sales and Operations Plan

FIGURE 4-12   Stock/Requirements List for NRG-A bars after disaggregation

The Stock/Requirements List is aptly named. It shows current stock, required materials, material expected to be received, and availability.

Exercise 4.3

Using the sales forecast for July through December from Exercise 4.2, develop a spreadsheet for sales and operations planning. Use the format of the spreadsheet shown in Figure 4-5. The number of working days for each month is shown in Figure 4-13.

	July	Aug.	Sept.	Oct.	Nov.	Dec.
Working Days	22	18	20	23	20	17

FIGURE 4-13 The number of working days at Fitter Snacker, July through December

Note that there are fewer working days in August because of an annual plant shutdown. The number of working days is also low in December because of the Christmas and New Year holidays.

For your production plan, try to keep the capacity utilization at 95 percent or less. To disaggregate the plan for the group into plans for NRG-A and NRG-B bars, use 70 percent of sales for NRG-A bars and 30 percent for NRG-B bars.

Demand Management

The Demand Management step of the production planning process links the Sales and Operations Planning process with the Detailed Scheduling and MRP processes. The output of the Demand Management process is the **master production schedule (MPS)**, which is the production plan for all finished goods. For Fitter Snacker, the MPS is an input to Detailed Scheduling, which determines which bars to make and when to make them. The MPS is also an input to the MRP process, which determines what raw materials to order to support the production schedule.

The Demand Management process splits FS's monthly production planning values into finer time periods. Figure 4-14 shows January's production plan by week and by day.

FS will use the weekly plan to plan materials management for Purchasing. Daily plans will be used for the product(s) that are to be produced on the snack bar line. The calculations were performed as follows:

For the weekly plan, the MPS plan for NRG-A bars in week 1 was calculated as:

4,134 cases in Jan. (monthly demand) × 4 working days in week 1 ÷ 22 working days in month of Jan. = 751.6 cases

This figure was rounded to 752 cases in Figure 4-14.

Because week 5 consists of the last three days in January and the first two days in February, the MPS for NRG-A bars in week 5 was calculated as:

4,134 cases in Jan. (monthly demand) × 3 working days in week 5 ÷ 22 working days in month of Jan. = 563.7 cases

	Week 1	Week 2	Week 3	Week 4	Week 5	
Demand Management	1/2 - 1/5	1/8 - 1/12	1/15 - 1/19	1/22 - 1/26	1/29 - 1/31	2/1 - 2/2
Monthly Demand NRG-A	4134	4134	4134	4134	4134	4198
NRG-B	1772	1772	1772	1772	1772	1799
Working Days in Week	4	5	5	5	3	2
Working Days in Month	22	22	22	22	22	20
MPS NRG-A	752	940	940	940	984	
Weekly Demand NRG-B	322	403	403	403	422	

Demand Management	Jan 2	Jan 3	Jan 4	Jan 5	Jan 6
Monthly Demand NRG-A	4134	4134	4134	4134	4134
NRG-B	1772	1772	1772	1772	1772
Working Days in Month	22	22	22	22	22
MPS NRG-A	188	188	188	188	188
Daily Demand NRG-B	81	81	81	81	81

FIGURE 4-14   Fitter Snacker's production plan for January: The first five weeks of production are followed by a day-by-day disaggregation of week 1

4,198 cases in Feb. (monthly demand) × 2 working days in week 5 ÷ 20 working days in month of Feb. = 419.8 cases

Total = 563.7 + 419.8 = 983.5 cases

MPS cases for week 5 were rounded to 984. In Figure 4-14, the daily production plan values for NRG-A and NRG-B bars were calculated by taking the monthly production plan value and dividing it by the number of working days in the month.

Notice that the Demand Management process for Fitter Snacker involves no user input. Input is from the Sales and Operations Planning step. The SAP R/3 software also uses information from the factory calendar (working days in week and month) to calculate the MPS.

Fitter Snacker does not do this sort of planning because it has no way to formally share sales forecast data between Marketing and Production. The company cannot relate its possible sales to its capacity and to the time available to make the product. Thus, FS cannot produce an accurate MPS.

The MPS is an input to the Detailed Scheduling and MRP processes. MRP is discussed next, and Detailed Scheduling is discussed after that.

Exercise 4.4

Develop a weekly production plan for July like the one for January shown in Figure 4-14. For the weekly sales periods, the last week will include three days in August. The factory calendar information is shown in Figure 4-15.

	Week 1	Week 2	Week 3	Week 4	Week 5	
Demand Management	7/2 - 7/6	7/9 - 7/13	7/16 - 7/20	7/23 - 7/27	7/30 - 7/31	8/1 - 8/3
Working Days in Week	4	5	5	5	2	3
Working Days in Month	22	22	22	22	22	18

FIGURE 4-15   Fitter Snacker's factory calendar for July

Materials Requirements Planning (MRP)

Materials Requirements Planning (MRP) is the process that determines the quantity and timing of the production or purchase of subassemblies and raw materials needed to support the MPS. The MRP process answers the question "What do we need to order to produce the product, and when do we need to order these raw materials?" In this section, you will see examples of how Fitter Snacker could accurately plan its raw materials purchases if it had ERP.

In Fitter Snacker's case, all product components (ingredients, snack bar wrappers, and display boxes) are purchased, so the company could use the MRP process to determine the timing and quantities for purchase orders. To understand MRP, you must understand the bill of material, the material's lead-time, and the material's lot sizing.

Bill of Material

The **bill of material (BOM)** is a list of the materials (including quantities) needed to make a product. The BOM for a 500-pound batch of the NRG-A or NRG-B bars is shown in Figure 4-16.

	Quantity	
Ingredient	NRG-A	NRG-B
Oats (lb.)	300	250
Wheat germ (lb.)	50	50
Cinnamon (lb.)	5	5
Nutmeg (lb.)	2	2
Cloves (lb.)	1	1
Honey (gal.)	10	10
Canola Oil (gal.)	7	7
Vit./Min. Powder (lb.)	5	5
Carob Chips (lb.)	50	
Raisins (lb.)	50	
Protein Powder (lb.)		50
Hazelnuts (lb.)		30
Dates (lb.)		70

FIGURE 4-16   The bill of material (BOM) for Fitter Snacker's NRG bars

The BOM for Fitter Snacker's NRG bars is fairly simple because all ingredients are mixed together to form the dough. To produce many other products, however, component parts are joined into subassemblies that are then joined to form the finished product. It is obviously more complicated to calculate the raw material requirements for products with more complex BOMs.

Lead Times and Lot Sizing

The BOM can be used to calculate how *much* of each raw material is required to produce a finished product. To determine the *timing* and *quantity* of purchase orders, however, requires information on **lead times** and **lot sizing**.

For example, if a manufacturer orders a make-to-stock item, the lead-time is the sum of the time required for the supplier to receive and process the order, take the material out of stock, package it, load it on a truck, and deliver it to the manufacturer. In addition, the manufacturer might include the time required to receive the material in its warehouse (unloading the truck, inspecting the goods, and moving the goods into a storage location).

Lot sizing refers to the process of determining production quantities (for raw materials produced "in house") and order quantities (for purchased items). In FS's case, many raw materials can only be ordered from a supplier in certain bulk quantities. For example, because FS uses large quantities of oats, the most cost-effective way to purchase oats is in bulk hopper-truck quantities, which means that the material must be ordered in 44,000-pound quantities. Wheat germ, however, is used in smaller quantities, and to avoid having wheat germ become stale, FS orders it in 2,000-pound bulk containers. Protein powder is packaged in 50-pound bags that are loaded 25 to a pallet, so the most cost-effective way to order protein powder is by the pallet load (1,250 pounds).

Let's look at the MRP process using oats, which have a two-week lead-time and must be ordered in hopper-truck quantities (multiples of 44,000 pounds). To determine when and how many pounds of oats should be ordered, we'll start with the weekly master production schedule (MPS) for NRG-A and NRG-B bars, and then:

1. Convert the quantities from cases to 500-pound batches.
2. Multiply the number of batches by the pounds-per-batch quantities (which is given in the BOM) to get the gross requirements for each raw material.
3. Subtract the existing raw material inventory and purchase orders that have already been placed from the gross requirements to determine the net requirements.
4. Plan orders in multiples of the 44,000-pound lot size, allowing for the two-week lead-time required for oats, to meet the net requirements in step 3.

These steps are summarized in Figure 4-17. This view of the data is frequently called an **MRP record**, which is the standard way of viewing the MRP process on paper.

Oats Lead Time = 2 weeks		Week 1	Week 2	Week 3	Week 4	Week 5
MPS	NRG-A	752	940	940	940	984
(cases)	NRG-B	322	403	403	403	422
MPS	NRG-A	108	135	135	135	142
(500 lb. batches)	NRG-B	46	58	58	58	61
Gross Requirements (lb)		44,090	55,087	55,087	55,087	57,667
Scheduled Receipts		44,000	44,000			
Planned Receipts				88,000	44,000	44,000
On Hand	11,650	11,560	473	33,386	22,299	8,632
Planned Orders		88,000	44,000	44,000		

FIGURE 4-17   The MRP record for NRG bars, weeks 1 through 5

The first two rows of the MRP record are the MPS that was the output from Demand Management (shown in Figure 4-14). These production quantities are in terms of shipping cases. The first step is to convert the MPS from shipping cases to 500-pound batches. Each shipping case weighs 72 pounds, so to convert shipping cases to 500-pound batches, multiply the number of shipping cases by 72 pounds per case, and then divide by 500 pounds per batch. Thus, producing 752 shipping cases of NRG-A bars in week 1 of the year will require 108 batches, as shown in Figure 4-17.

The next row in Figure 4-17 is Gross Requirements. The Gross Requirements figures are calculated by multiplying the MPS quantity (in production batches) by the pounds of oats needed for a batch of snack bars. FS uses 300 pounds of oats per batch of NRG-A bar and 250 pounds of oats per batch of NRG-B bar. This information is derived from the BOM (Figure 4-16). Therefore, for week 1 FS needs:

NRG-A: 108.3 batches × 300 lb. per batch = 32,490 lb. oats
NRG-B: 46.4 batches × 250 lb. per batch = 11,600 lb. oats
_____
Total = 44,090 lb. oats

The next row in Figure 4-17 is the Scheduled Receipts. This row shows the expected arrival dates of orders of materials that have already been placed, meaning that the supplier has been given the purchase order and is in the process of fulfilling it. There is a two-week lead-time for oats, so for oats to be available in week 1 and week 2 of the year, oats orders must be placed in the last two weeks of the previous year.

The next row, Planned Receipts, shows when Planned Orders will arrive. The Planned Receipts row is directly related to the Planned Orders row at the bottom of the record. A Planned Order is one that has not been placed with the supplier but will need to be placed to avoid running out of materials. Because there is a two-week lead-time for oat orders, the items in the Planned Orders will be available for production in two weeks, which is indicated by an entry in the Planned Receipts row. The arrows in Figure 4-17 show the relationship between Planned Orders and Planned Receipts. For example, the Planned Order for 88,000 pounds of oats in week 1 will be available for use in week 3, which is shown by the Planned Receipt of 88,000 pounds in week 3. The MRP calculation suggests that an order for 88,000 pounds of oats should be placed in week 1 so that it arrives in week 3; there is only one order, but it shows up in two places on the MRP record.

The next row in Figure 4-17 is the On Hand row. The first number in this row (11,650) is the inventory of oats on hand at the beginning of week 1. The number in the week 1 column (11,560) is a projection of the inventory that will be on hand at the end of week 1 (and therefore at the beginning of week 2)—accounting for the beginning inventory, gross requirements, and planned and scheduled receipts. In the case of the on-hand value for week 1, the initial inventory of 11,650 pounds, plus the 44,000 pounds of scheduled delivery, minus the 44,090 gross requirement leaves 11,560 pounds of oats available at the start of week 2.

The last row is the Planned Orders row. This is the quantity that the MRP calculation recommends ordering, and it is the output from the MRP process that is used by Purchasing to determine what to order to produce the product and when to order it.

Many times, a planner may intervene to tell the system to adjust the planned order. For example, notice that the on-hand quantity of oats in week 2 is only 473 pounds. This means that at the start of week 2, there will only be enough oats to mix one batch of dough. Since the production line produces 3,000 pounds of bars an hour, one batch of dough will keep the production line running for only 10 minutes. If the scheduled order does not arrive early enough on the first day of week 2, the production line could be shut down. When the purchase order scheduled to arrive in week 1 was ready to be placed (two weeks prior to the beginning of week 1), the planner should have evaluated that order, considering the low inventory level projected for the beginning of week 2. The planner might have decided to place an order for two hopper-truck loads of oats, instead of the planned order for one load. Or he could have ensured that the scheduled receipt shown in week 2 would actually be delivered at the end of week 1. Planning factors like lead-times are just estimates, and so someone should evaluate the planned orders suggested by the MRP calculation before allowing the program to automatically turn them into purchase orders.

Notice once again the need for software to help with this kind of calculation. Of course, a human can do these computations, but with many products and constituent materials, the calculations become very tedious and are prone to error. Even for a small company such as FS, doing the calculations with reasonable speed and accuracy requires software help. Notice also the information needed to do the MRP calculation: Starting with a sales forecast, the software works down to the MPS and then to a schedule of needed raw materials.

Exercise 4.5

Develop an MRP record, similar to the one in Figure 4-17, for wheat germ for the five weeks of January. Wheat germ must be ordered in bulk container quantities, so the planned orders must be in multiples of 2,000 pounds. Use a lead-time of one week, an initial on-hand inventory of 1,184 pounds, and assume that an order of 8,000 pounds is scheduled for receipt during week 1. Are there any weeks when you, as a planner, would consider placing an order above or below the minimum required? Why? Assume that there are no problems with storage capacity or shelf life.

Exercise 4.6

Fitter Snacker's purchasing policy has been to carry high levels of inventory to avoid stockouts. Why can inventory levels be lower with an integrated ERP system and MRP-based purchasing? If you had to calculate the financial advantage of this change, how would you do it?

SAP R/3 MRP and MRP Records

The MRP list in SAP R/3 looks very much like a Stock/Requirements List, which you already saw in Figure 4-12. The MRP list shows the results of the MRP calculations, while the Stock/Requirements List shows those results plus any changes that have occurred since the MRP list was generated (planned orders converted to purchase orders or production orders, material receipts, and so on). Because the MRP calculations are time consuming to process for a company producing hundreds of products using thousands of parts, the MRP process is usually repeated every few days—or perhaps even weekly. The Stock/Requirements List allows the users of the system to see what is happening (and will happen) with a material in real time. Compare the data shown in the MRP record in Figure 4-17 with the MRP list in Figure 4-18 and the Stock/Requirements List in Figure 4-19.

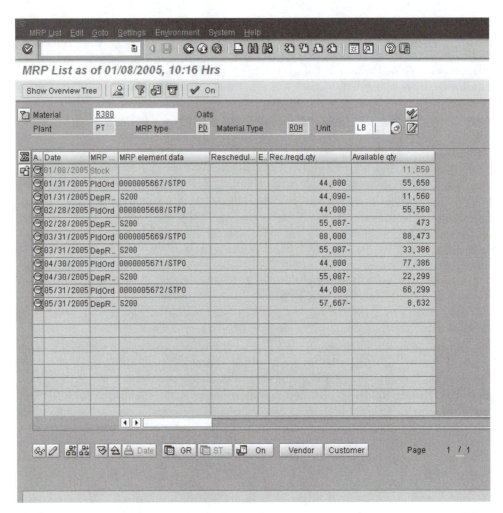

FIGURE 4-18 The MRP List in SAP R/3

The MRP list in Figure 4-18 shows Planned Orders (PldOrd) and Dependent Requirements (DepR) in the second column. Dependent Requirements represent the demand for
snack bar dough from Planned Orders. Each batch of snack bar dough planned by the system requires 300 pounds of oats for NRG-A bars and 250 pounds of oats for NRG-B bars,
which are referred to as Dependent Requirements because they are created by the production plans for snack bar dough. The MRP process creates Planned Orders to meet these
dependent requirements. The Planned Orders are recommendations by the system to create orders (in this case, Purchase Orders) for oats.

FIGURE 4-19 The Stock/Requirements list in SAP R/3

The Stock/Requirements list shown in Figure 4-19 shows Planned Orders, but it also shows Purchase Requisitions (PurRqs) and Purchase Orders (POitem). When a planner decides that it is time for a Planned Order to become a Purchase Order, the Planned Order is converted to a Purchase Requisition, which is a request to Purchasing to create a purchase order. The planner can convert a planned order to a purchase order from the Stock/ Requirements list by double-clicking the Planned Order line. This action calls up the window shown in Figure 4-20.

From this window, the planner can create a Purchase Requisition or review the Planned Order and make changes before creating the requisition. SAP R/3 also provides the ability to mass process planned orders, which converts groups of Planned Orders to Purchase Orders simultaneously. The MRP process can also be configured to automatically create Purchase Requisitions; for example, all Planned Orders created within one week of the MRP calculation could be automatically created as Purchase Requisitions.

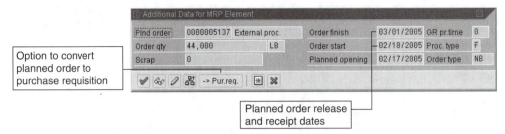

Option to convert planned order to purchase requisition

Planned order release and receipt dates

FIGURE 4-20 Conversion of Planned Order to Purchase Requisition

Once a Purchase Requisition is created, an employee in the Purchasing department has to turn it into a Purchase Order. One of the important steps in creating a purchase order is choosing the best vendor to supply the material. An integrated information system like SAP can facilitate this process. Figure 4-21 shows the Source Overview screen, which provides access to information that can help to select the vendor. From this screen, the purchasing employee can view information about each vendor, simulate the price from each vendor, which might include quantity discounts and transportation costs, and also look at the vendor evaluation—the company's rating of the vendor. The SAP R/3 system can be configured to rate vendors based on a number of performance criteria, including quality of goods provided and on-time delivery. The evaluation scores for each vendor are updated automatically as materials are received. The integrated information system allows purchasing to make the best decision on a vendor based on relevant, up-to-date information.

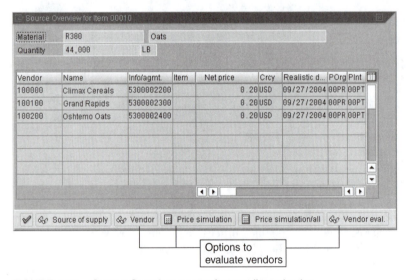

Options to evaluate vendors

FIGURE 4-21 Source Overview screen for supplier selection

Once the purchasing employee decides which vendor to use, the Purchase Order is transmitted to the vendor. The SAP R/3 system can print out a paper order that can be mailed to the vendor. More likely, the system will be configured to either fax the order to the vendor or transmit it electronically through EDI (electronic data interchange) or over the Internet.

Detailed Scheduling

Finally, let's examine the last portion of the production process, Detailed Scheduling. The aggregate production plan for product groups developed in Sales and Operations planning has been disaggregated to individual products in finer time increments through the demand management process. In Detailed Scheduling, a detailed plan of what is to be produced needs to be developed, considering machine capacity and available labor. Detailed Scheduling is somewhat complex and tedious and will not be presented here in spreadsheet form, but the important concepts and issues are discussed. A key decision in detailed production scheduling is how long to make the production runs for each product. Longer production runs mean that fewer machine setups are required, reducing the production costs and increasing the effective capacity of the equipment. On the other hand, shorter production runs can be used to lower the inventory levels for finished products. Thus, the production run length requires a balance between setup costs and holding costs to minimize total costs to the company.

Because the capacity of the Fitter Snacker production mixers is much greater than that of the snack bar production line, scheduling mixer production is not an issue. Because the dough must be mixed before the snack bar production line can start, employees who run the mixers at FS should begin working a half-hour before the employees who run the production line. Therefore, four batches of dough can be mixed before the production line starts. With a bit of a head start and a detailed schedule for the production line, it is a simple matter for the personnel operating the mixers to keep ahead of the production line. Thus, the key step is to develop a detailed production schedule for the snack bar production line.

The manufacturing process that Fitter Snacker uses is known as **repetitive manufacturing**. Repetitive manufacturing environments usually involve production lines that are switched from one product to another similar product. Most consumer packaged goods are produced in repetitive manufacturing environments. In repetitive manufacturing, production lines are scheduled for a period of time, rather than for a specific number of items, although it is possible to estimate the number of items that will be produced over a period of time. For Fitter Snacker, the production schedule for a week might be to produce NRG-A bars from Monday morning until the end of Wednesday, then change over to NRG-B bars for all of Thursday until Friday at noon, when the production line will be switched back to NRG-A bars. Given this schedule, it is possible to estimate the number of bars that will be produced. Figure 4-22 shows the repetitive manufacturing planning screen in the SAP R/3 system. This screen allows the planner to view capacity, production schedule length and quantity produced in one screen.

In some companies, responsibility for inventory costs are assigned to a Materials Management group, and capacity utilization performance is assigned to a Production group. The Materials Management group wants short production runs to keep inventory levels down, while the Production group wants long production runs to keep capacity utilization high. Instead of a decision in which total costs are minimized for the benefit of the company, the decision regarding production run length can be a source of organizational bickering.

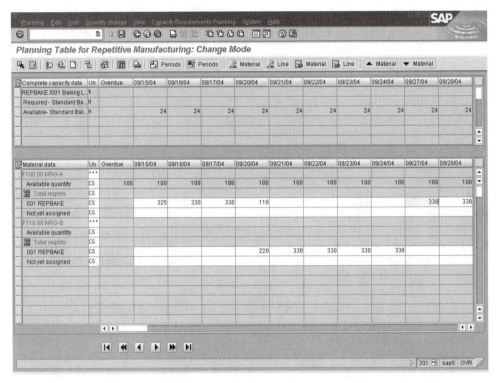

FIGURE 4-22 Repetitive manufacturing planning table in SAP R/3

This conflict points out an advantage of production planning in an ERP system. Because the goal of the company is to maximize profit, the duration of production runs should be decided by evaluating the cost of equipment setup and holding inventory. This analysis is simplified in an integrated information system, in which accounting information is collected automatically. This accounting information allows managers to better evaluate schedule trade-offs in terms of costs to the company.

Providing Production Data to Accounting

In Chapter 1 you learned that functional areas must share data for a company to be successful. Accounting needs to know what Manufacturing has produced and what resources were used in producing those products to determine which products, if any, are producing a profit—and then provide information for management to determine how to increase profits. In the manufacturing plant, ERP packages do not directly connect with production machines. For example, in Fitter Snacker's case, SAP R/3 could not directly read the number of bars that came off the packing segment of the snack bar line (Figure 4-1). The data must be gathered in some way and then entered into SAP R/3 for inventory accounting purposes.

Data can be entered into SAP R/3 through a PC on the shop floor, scanned by a barcode reader, or entered using a wireless PDA. SAP R/3 is an open-architecture system, meaning that automated data-collection tools marketed by third-party hardware and software companies can be integrated with SAP R/3. Soon, Radio frequency identification (RFID) technology will simplify this process. RFID technology is discussed in more detail in Chapter 8.

In an integrated ERP system, the accounting impact of a material transaction can be recorded automatically. For example, when a shipment of oats arrives at the Fitter Snacker plant, someone in the Receiving department must verify the material and the quantity and quality of the shipment before it is accepted. Once the material is accepted, the fact that the material has arrived and has been accepted must be communicated to the SAP R/3 system. This is done by completing a goods receipt transaction, which is shown in Figure 4-23.

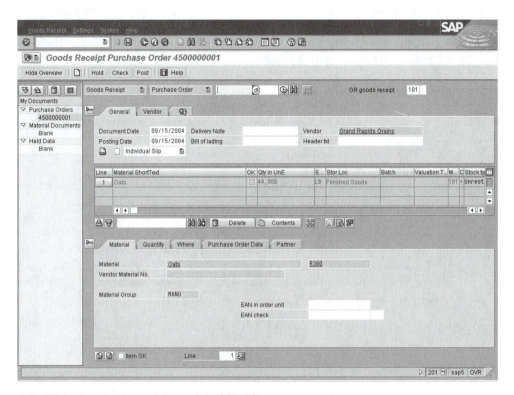

FIGURE 4-23   Goods receipt screen in SAP R/3

The goods receipt must be matched with the Purchase Order that initiated it to make sure that exact materials ordered have been received so that the vendor can be paid. It is possible for the quantity of material entered in the goods receipt to differ from the quantity specified on the purchase order. Depending on the configuration settings, the SAP R/3 system might block entry of the receipt if the discrepancy is too large. If the discrepancy is small, then the receipt may be allowed, with the difference posted to the correct variance account, which allows the transaction to be processed but maintains a record for management to review to see if there is a consistent problem with a vendor "shorting" an order (consistently shipping less than was ordered).

When the receipt is successfully recorded, the SAP R/3 system immediately records the increase in inventory levels for the material. On the accounting side of the system, the value of the inventory shown in the general ledger account will be automatically increased as well. This is an important feature of an integrated information system: the goods receipt is recorded once, but the information is immediately available to both manufacturing and accounting—and the information is consistent. An integrated information system also has the ability to adjust for changes in material costs. If cost of the material changes frequently, the system can be configured to reevaluate the value of all of the inventory of that material that the company has. For example, suppose Fitter Snacker has 10,000 pounds of cinnamon that it bought at $3.00 per pound. The company would value its inventory of cinnamon at $30,000. Suppose the price of cinnamon has risen to $4.00 per pound and Fitter Snacker just purchased 1,000 pounds of cinnamon at the higher price. What is the value of the 11,000 pounds of cinnamon that Fitter Snacker now owns? The SAP R/3 system can be configured to use a moving average formula to reevaluate the inventory based on the current market prices. Depending on the exact nature of the formula, the cinnamon would be valued at somewhere between the $3.00 per pound previously assumed and the $4.00 per pound that was just paid. This calculation can be performed automatically each time material is received.

Using an ERP package to record data does not necessarily make the shop-floor accounting data more accurate. The ERP system allows data to be entered in real time. Furthermore, capturing data for manufacturing and inventory purposes on the shop floor means that it is captured at the same time for accounting and inventory management purposes, which eliminates any need to reconcile accounting and manufacturing records. But it requires employees to follow the process. If employees can take material out of inventory without recording the transaction, then the real-time information in the ERP system is worthless. Technologies like bar-code scanners and RFID tags can help in this process, but accurate data require both a capable information system and properly trained and motivated employees.

Exercise 4.7

Briefly describe how the implementation of SAP R/3 might change the relationship between Production and Warehousing at Fitter Snacker.

ERP AND SUPPLIERS

As mentioned in the introduction to this chapter, Fitter Snacker is part of a supply chain that starts with farmers growing oats and wheat and ends with a customer buying an NRG bar from a retail store. Previously, companies have used competitive bidding to achieve low prices from suppliers. This frequently led to an adversarial relationship between suppliers and customers. In recent years, companies have realized that they are part of a supply chain, and if the supply chain is more efficient, all participants in the chain can benefit. Collaboration can frequently achieve more than competition, and ERP systems can play a key role in collaborative planning.

Working with suppliers in a collaborative fashion requires trust among all parties. A company opens its records to its suppliers, and suppliers can read the company's data because of common data formats. Working this way with suppliers cuts down on paper-work and response times. Reductions in paperwork, savings in time, and other efficiency improvements translate into cost savings for the company and the suppliers. ERP lets companies and suppliers share information (sales, inventory, production plans, and so on) in real time throughout the supply chain. This allows all parties (suppliers, manufacturers, and customers) to drive costs that don't add value to the product (such as inventory, overtime, changeovers, and spoilage) out of the supply chain while simultaneously improving customer service.

The Traditional Supply Chain

In a traditional supply chain, information is passed through the supply chain reactively as participants increase their product orders—as illustrated in Figure 4-24. For example, a retailer sees an increase in the sales of FS's bars and orders a larger quantity of bars from the wholesaler. If a number of retailers increase their orders, the wholesaler will increase its orders from Fitter Snacker. When FS gets larger orders from wholesalers, it must increase production to meet the increased demand. To increase production, FS will order more raw materials from suppliers.

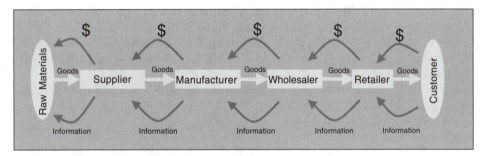

FIGURE 4-24   Supply chain management (SCM) from raw materials to consumer

Because of the time lags inherent in a traditional supply chain, it might take weeks—or even months—for information about FS's increased need for raw materials to reach FS's suppliers. Raw material suppliers may require time to increase *their* production to meet FS's larger orders, resulting in temporary shortages for the supplier.

By contrast, if the participants in the supply chain are part of an integrated process, information about the increased customer demand can be passed quickly through the supply chain, so each link in the chain can react quickly to the change.

EDI and ERP

The development of supply-chain strategies does not necessarily require an ERP system. Before ERP systems were available, companies could be linked with customers and suppliers through electronic data interchange (EDI) systems. Recall from Chapter 2 that EDI is the computer-to-computer exchange of standard business documents (such as purchase orders) between two companies. A well-developed ERP system, however, can facilitate SCM because the needed production planning and purchasing systems are already in place. In addition, the integration of accounting data in the ERP system (described in the next chapter) allows management to evaluate changes in the market and make decisions about how those changes should affect the production plan. With an ERP system, sharing production plans along the supply chain can occur in real time. Using the Internet can make this communication even faster and cheaper than if done by private EDI networks.

The Measures of Success

Performance measurements (sometimes referred to as **metrics**) have been developed to show the effects of better supply-chain management. One measure is called the **cash-to-cash cycle time**. This refers to the time between paying for raw materials and collecting cash from the customer. In one study, companies with efficient SCM processes do this in a month, whereas those without SCM take 100 days.[1]

Another measure is total SCM costs. These costs include the cost of buying and handling inventory, processing orders, and information systems support. In one study, companies with efficient SCM processes incur costs equal to 5 percent of sales. By contrast, companies without SCM incur costs of up to 12 percent of sales.[2]

Other metrics have been developed to measure what is happening between a company and its suppliers. For example, Staples, the office-supply company, measures three facets of the relationship.[3] **Initial fill rate** is the percentage of the order that the supplier provided in the first shipment. Another metric is **initial order lead time**, which is the time needed for the supplier to fill the order. Finally, Staples measures **on-time performance**. If the supplier agreed to requested delivery dates, this measurement tracks how often the supplier actually met those dates.

Improvement in metrics like these leads to improvement in overall supply chain cost measurements.

Exercise 4.8

Assume a manufacturer of residential lawn and garden equipment is considering investing in hardware and software that will improve linkages with suppliers. Management expects to save 5 percent of sales by tightening up the supply chain in the first year, 3 percent in the second year, and 1 percent in the third year. The company's annual sales are $1 billion. The company's chief financial officer insists that the investment must pay for itself in cost savings in three years. To meet this requirement, how much should the chief information officer be allowed to spend on improving the supply chain? Explain your answer.

ANOTHER LOOK

SCM and its Critical Success Factors

For a supply chain management project to be successful, the company must achieve certain key factors. The Advisory Council at *Information Week* has put together a list of those key factors:[4]

- *Business-driven strategy.* The information system for managing the supply chain must focus on the customer, allowing the customer greater efficiencies in the process. The customer should find that ordering material from the vendor is now more efficient with the new system. The system is based on a complex business process model.
- *Management commitment.* As with any IT project, top management must be committed to support the project; otherwise the project is doomed to fail.
- *Supplier and partner expectation management.* An SCM implementation requires change from all parties. A company implementing an SCM project must be prepared to work closely and manage that change from all angles.
- *Internal expectation management.* Change management is especially important within in a company. Personnel involved in a new system must be informed and trained to manage any anxiety they might have in their job changes.
- *Learning period acceptance.* Companies implementing SCM must accept that it takes a long time to fully realize the benefits of the system.

In addition, SCM has the benefit of driving business value. Simply put, SCM will help your company's bottom line, because if you make it more efficient for customers to buy your product and you make it more efficient to buy your suppliers' products, then your overall company will benefit and become more efficient.

Implementing an SCM system without paying attention to the critical factors listed above will result in poor performance. In fact, a large percentage of companies that implement SCM are not satisfied. Forty-five percent of respondents to a Booz Allen Hamilton supply-chain management survey report that their IT solutions are failing to meet expectations. SCM consultants often find that companies have unrealistic expectations of the technology. Companies must decide how their supply chains should be designed and run, and then use technology to implement those goals.[5]

Questions:

1. Assume you were recently hired in an IT role at a large manufacturing company. Your first job is to convince the manufacturing manager that the implementation of an SCM system is feasible. Write a memo to that manager highlighting advice on how to ensure the smooth implementation of such a system.
2. Outline the potential pitfalls that your company may encounter when implementing SCM.

Chapter Summary

1. An ERP system can improve the efficiency of production and purchasing processes. Efficiency begins with Marketing sharing a sales forecast. A production plan is created based on that forecast and shared with Purchasing so raw materials can be ordered properly.

2. Production planning can be done without an ERP system, but an ERP system that contains MRP allows Production to be linked to Purchasing and Accounting. This data sharing increases a company's overall efficiency.

3. Companies are building on their ERP systems and integrated systems philosophy to practice SCM. In doing this, the company looks at itself as part of a larger process that includes customers and suppliers. By using information more efficiently along the entire chain, significant cost savings can result.

Key Terms

Bill of material (BOM)

Cash-to-cash cycle time

Initial fill rate

Initial order lead time

Lead time

Lot sizing

Master production schedule (MPS)

Metrics

MRP record

On-time performance

Repetitive manufacturing

Rough-cut planning

Standard costs

Exercises

1. Review the information in this chapter regarding Fitter Snacker's production problems and create a flowchart to document that flawed process. Use the flowcharting tools in Microsoft Excel to produce a professional looking diagram. Include a paragraph about how ERP systems will help remove flaws from the FS production process.

2. Using an Internet search engine or the library, research a manufacturing company that has successfully implemented SCM. Did the company have an ERP system prior to the SCM implementation? Can the company quantify the benefits of having SCM? How does the company measure success?

3. Summarize the sources and destinations of production and purchasing information within a company that has MRP and ERP. Refer to Figure 4-2 as a guide. Data could be summarized in a tabular format using the table feature in a word processor. Use three columns: Data, Source, and Provided to.

4. Interview a professor at your university who teaches in the operations research area. Ask the professor about his or her industry contacts and how those contacts use ERP, MRP, and SCM. Ask the professor for the name of a production manager in a manufacturing firm, and then try to get an appointment with that manager. If possible, interview that production manager and ask about the flow and channels of information to and from the production department.

5. Compare customer relationship management (CRM) and SCM. How are they similar? How are they different? In answering, consider the kinds of technologies used in each.

End Notes

[1] Christopher Koch, "The Big Payoff," *CIO* (October 1, 2000) (archived at *http://www.cio.com/archive/101000_payoff.html*).

[2] Koch, "The Big Payoff."

[3] Koch, "The Big Payoff."

[4] "Smart Advice: Top-Down Strategy," *Information Week* (February 23, 2004).

[5] "Supply Chain Management Still A Work In Progress," *Information Week* (May 23, 2003).

CHAPTER **5**

ACCOUNTING IN ERP SYSTEMS

LEARNING OBJECTIVES

After completing this chapter, you should be able to:

- Describe the differences between financial and managerial accounting.
- Identify and describe problems associated with accounting and financial reporting in unintegrated information systems.
- Describe how ERP systems can help solve accounting and financial reporting problems in an unintegrated system.
- Describe how the Enron scandal and the Sarbanes-Oxley Act will affect accounting information systems.
- Explain accounting and management reporting benefits that accrue from having an ERP system.

INTRODUCTION

In previous chapters, you read about functional area activities, both generally and specifically: In Chapter 1, you read an overview of functional area activities; in Chapter 3, you read about Marketing and Sales activities; and in Chapter 4, you read about Supply Chain Management. In this chapter, you'll read about the activities in another functional area, Accounting. You will see how Accounting is tightly integrated with all of the other functional areas, as well as how Accounting activities are necessary for decision making.

ACCOUNTING ACTIVITIES

Accounting activities can generally be classified as either Financial Accounting or Managerial Accounting. An additional area of accounting, Tax Accounting, is beyond the scope of this text. Because Tax Accounting is chiefly external reporting of a business's activities to the Internal Revenue Service, data gathered for Financial Accounting forms the basis for tax accounting.

Financial Accounting consists of documenting all transactions of a company that have an impact on the financial state of the firm, and then using these documented transactions to create reports to external parties and agencies. These reports, or financial statements, must follow prescribed rules and guidelines of various agencies, such as the Financial Accounting Standards Board (FASB), the Securities and Exchange Commission (SEC), and the Internal Revenue Service (IRS).

Common financial statements include balance sheets and income statements. The **balance sheet** is a statement that shows account balances such as cash held, amounts owed to the company by customers, the cost of raw materials and finished-goods inventory, long-term assets such as buildings, amounts owed to vendors, amounts owed to banks and other creditors, and amounts that the owners have invested in the company. A balance sheet is a good overview of a company's financial health at a point in time, a key consideration for a company's creditors and owners. Figure 5-1 shows a balance sheet for Fitter Snacker.

Fitter Snacker Balance Sheet At December 31, 2005 (in thousands of dollars)		
Assets		
Cash		$5,003
Accounts receivable		$4,715
Inventories		$9,025
Plant and equipment		$6,231
Land		$1,142
Total assets		$26,116
Liabilities		
Accounts payable	$6,400	
Notes payable	$10,000	
Total liabilities		$16,400
Stockholders' Equity		
Contributed capital	$2,000	
Retained earnings	$7,716	
Total stockholders' equity		$9,716
Total liabilities and stockholders' equity		$26,116

FIGURE 5-1 Fitter Snacker sample balance sheet

The **income statement** shows the company's sales, cost of sales, and the profit or loss for a period of time (typically quarter or year). Profitability is important to creditors and owners. It is also important information for managers in charge of day-to-day operations. A manager sees profits as indicators of success and losses as indicators of problems to be solved. You can get an idea of the contents of an income statement by looking at Figure 5-2, which shows an income statement for Fitter Snacker.

Fitter Snacker Income Statement For the Year Ended December 31, 2005 (in thousands of dollars)		
Revenues		
Sales revenue	$36,002	
Total revenues		$36,002
Expenses		
Cost of goods sold expense	$25,691	
Selling, general, and administrative expense	$4,251	
Research and development expense	$962	
Interest expense	$521	
Total expenses		$31,425
Pretax income		$4,577
Income tax expense		$1,144
Net income		$3,433

FIGURE 5-2 Fitter Snacker sample income statement

Companies prepare financial statements at least quarterly, and to prepare these statements companies must "close their books." In accounting, the closing process means that temporary or nominal accounts (such as revenue, expense, gain and loss) have their balances transferred or closed to the Retained Earnings Account. Closing means that these nominal accounts will have a zero balance from which to start accumulating revenues and expenses in the next reporting period. Closing entries are made to transfer balances and to establish a zero balance. To do this, the accounts must be checked to see that they are accurate and up to date. If a company's information systems routinely generate accurate and timely data, closing the books can go smoothly. If they are not, "adjusting" entries must be made, and, in that case, closing the books can be a very time-consuming chore with innaccurate results.

Managerial Accounting deals with determining the costs and profitability of the company's activities. While the information in a company's balance sheet and income statement show whether a firm is making an overall profit, the goal of managerial accounting is to provide managers with detailed information so that the profitability of a particular product, sales region, or marketing campaign can be determined. Managerial accounting provides information managers use to control a company's day-to-day activities. Managerial accounting also helps managers to develop long-term plans for operations, marketing, personnel needs, repayment of debt, and other management issues. Because managerial accounting provides reports and analyses for internal use, companies have great flexibility in how they configure their managerial accounting systems.

Using ERP for Accounting Information

Recall from Chapter 1 that in the past, companies have had separate functional information systems: a marketing information system, a manufacturing information system, and so on, each with its own way of gathering data and its own file system for recording data. Such unintegrated systems were built primarily to handle the needs of the functional area and secondarily to provide data to Accounting. With unintegrated systems, the functional areas shared their data with Accounting, so Accounting could "keep the books," that is, keep a record of all financial transactions. Data sharing, however, usually did not occur in real time, so Accounting's data were not up to date. In addition, the shared data might not be the only information that Accounting needed to prepare reports for management. Accountants and functional area clerks had to do significant research to prepare reports. Since the 1960s, legions of accountants, analysts, and programmers have tried to make unintegrated systems work. This approach has not worked very well, as prior chapters illustrate.

An ERP system, with its centralized database, avoids these problems. For example, suppose finished goods are transferred from the assembly line to the warehouse. The transaction can be easily recorded in the warehouse by using a terminal or a bar code scanner. In SAP R/3, for example, the Materials Management module would see the transfer event as an increase in finished-goods inventory available for shipment; the Accounting module would see the event as an increase in the monetary value of finished goods. With ERP, everyone uses the same database to record operating data. This database is then used to generate management reports, make financial statements, and create budgets.

In traditional accounting, a company's accounts are kept in a record called the **general ledger**. In the SAP R/3 system, input to the general ledger occurs simultaneously with the business transaction in the specific module. Many SAP R/3 modules cause transaction data to be entered into the general ledger, including:

- Sales and Distribution (SD), which lets the user record sales and creates an accounts receivable entry (a general ledger document that indicates a customer owes money for the goods received).
- Materials Management (MM), which controls purchasing and recording of inventory changes. When a purchase order is created, an accounts payable entry is created in the general ledger noting that the company has an obligation to pay for goods that will be received. Whenever material is moved into

or out of inventory (purchased materials are received from the vendor, materials are taken out of inventory to support production, or finished goods are taken from production and put into inventory), general ledger accounts are affected.

- Financial Accounting (FI), which manages the accounts receivable and accounts payable items created in the SD and MM modules, respectively. The FI module is also where the general ledger accounts are closed at the end of a fiscal period (quarter or year) and financial statements are generated.
- Controlling (CO), which tracks the costs associated with producing products. To make a profit, it is critical for the company to have an accurate picture of its product costs to make correct decisions about product pricing and promotions as well as capital investments.
- Human Resources (HR), which manages the recruiting, hiring, compensation, termination, and severance of employees. The HR module also manages benefits and generates the payroll.
- Asset Management (AM), which is used to manage fixed-asset purchases (plant and machinery) and the related depreciation.

OPERATIONAL DECISION-MAKING PROBLEM: CREDIT MANAGEMENT

A company with an unintegrated information system can have accounting data that is out-of-date or inaccurate. Out-of-date data can cause problems when the company is trying to make operational decisions. This problem was illustrated in Chapter 3 by Fitter Snacker's problems in making credit decisions. In this section, FS's credit-granting problems will be discussed in more detail. First, we'll look at industrial credit granting in general, and then at FS's credit-check problem.

Industrial Credit Management

Companies routinely sell to customers on credit. Good financial management requires that only so much credit be extended to a customer. At some point, the customer must pay off some of the debt to justify the faith the seller has shown (and so the seller can turn the accounts receivables into cash). Credit management requires a good balance between granting sufficient credit to support sales while making sure that the company does not lose too much money by granting credit to customers who end up defaulting on their credit obligations.

In practice, sellers manage this relationship by setting a limit on how much a customer can owe at any one time and then monitoring that limit as orders come in and payments are received. For example, the seller might tell a buyer that her credit limit is $10,000, which means that the most she can owe is $10,000. If the buyer reaches that amount, no further sales orders will be accepted until the buyer pays off some of her debt. When a sale is made on credit, the seller makes an entry on the books to increase his accounts receivable and his sales. Thus, when the buyer's accounts receivable balance on the seller's books reaches $10,000, some payment must be made to the seller.

Continuing the example, assume that the buyer calls the seller to order $3,000 of goods, and her credit limit is $10,000. If the seller's accounts receivable for the buyer is already $8,000, then the seller should not accept the $3,000 order because the accounts receivable balance would then be $11,000, which exceeds her credit limit. Instead of refusing the order, the seller's sales representative might suggest that the buyer reduce the size of the order, or ask her to send in a payment before processing the order, thus reducing the buyer's debt. Clearly, to make this system work, a sales representative needs to be able to access an up-to-date accounts receivable balance when an order comes in.

If Marketing and Accounting have unintegrated information systems, full cooperation between the two functional areas will not be easy to achieve. Marketing knows the current order's value, but Accounting keeps the accounts receivable records. If Accounting keeps the books up–to-date and can provide the current accounts receivable balance to Marketing *when needed*, then credit limits can be properly managed. Marketing can compare the customer's credit limit to an accurate balance-owed amount (plus the order's value) to make a decision. However, in an unintegrated system, Accounting may not immediately record sales and/or payment receipts as they occur. In that case, accounts receivable balances will not be current. Furthermore, the sales clerk may be working from an out-of-date credit-balance printout. If the printout balances do not reflect recent payments, a customer may be improperly denied credit. The customer would probably challenge the denial, which would trigger a request for updated information in Accounting. The delay entailed in that research could reduce customer satisfaction, and performing the research would consume valuable employee time.

These problems should not arise with an integrated information system. When a sale is made, accounts receivable is immediately increased. When a payment is received and recorded, accounts receivable is immediately decreased. Because the underlying database is available to Marketing *and* Accounting, sales representatives can also see customer records immediately. Thus, sales representatives do not need to make a request to Accounting for the customer's accounts receivable balance.

With that background, we can now consider Fitter Snacker's credit management.

Fitter Snacker's Credit Management Procedures

As described in Chapter 3, an FS sales clerk refers to a weekly printout of the customer's current balance and credit limit to see if credit should be granted. Assuming that the customer's order would not present credit-limit problems, the sales clerk enters the sale in the sales order-entry system, which is a stand-alone computer program. Sales data are transferred to Accounting by disk three times a week. An accounting clerk can use the sales input to prepare a customer invoice.

Accounting must make any adjustments for partial shipments before preparing the invoice. The accuracy of the adjustment process depends on whether the warehouse transmits order changes in a timely fashion. Accounting also makes the standard revenue-recognition accounting entry: debit accounts receivable and credit sales for the amount billed.

Accounting clerks also process customer payments. Checks are received and manually handled by the clerk. The data are entered in the accounting program, increasing cash and decreasing accounts receivable. The data are later used to make entries to individual customer accounts, reducing the amount owed to FS by customers. If time permits, accounts are posted on the day payment is received and the bank deposit is made; otherwise, the entries are done as soon as possible the next day. Thus, there can be some delay between the time FS receives a check from the customer and when the customer's accounts receivable balance is actually reduced.

Now let's look at how SAP R/3 could help FS's credit management.

Credit Management in SAP R/3

Suppose FS is using SAP R/3 as its ERP system. This system would allow FS to set a credit limit for each customer. Figure 5-3 shows the SAP R/3 screen where credit management is configured. A company can configure any number of credit check options in the SAP R/3 system. The figure shows a dynamic credit check with Reaction C selected. Reaction C means that if an order is being saved that will cause the customer to exceed its credit limit, the system will issue a warning indicating the amount by which the order exceeds the credit limit. Because the system is issuing a warning, the order can be saved, but will be blocked from further processing until the credit problem is cleared. Frequently, companies do not configure the system to provide warnings to sales order clerks because they are not equipped to correct the problem and because the credit problem is an issue between the selling firm's accounts receivable department and the customer's accounts payable department.

Figure 5-3 also shows that the credit check is Dynamic and has a two-month horizon. This means that only the next two months of sales orders will be used in calculating the credit check. Customers may place orders for a long-range schedule, but only those that will be shipped in the near term are usually considered in the credit check.

FIGURE 5-3 Credit management configuration

Figure 5-4 shows the credit checking process in Figure 5-3 applied to a specific customer, Health Express. Health Express has a credit limit of $1,000 and currently has used $590 of this limit. If Health Express places an order for snack bars that totals more than $410, then the order will be blocked.

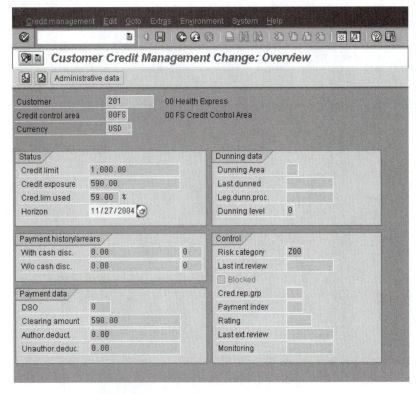

FIGURE 5-4 Credit management for Health Express

Figure 5-5 shows the transaction where blocked sales orders are listed. Most companies have an employee who is responsible for reviewing blocked sales orders (perhaps every two hours) and taking corrective action. The advantage of using the SAP R/3 system to manage credit is that the process is automated and the data are available in real time. The user can double-click the sales order to see company information such as contacts or to see payment history.

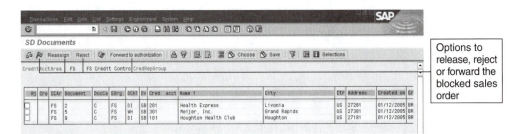

Options to release, reject or forward the blocked sales order

FIGURE 5-5 Blocked sales order

116

Chapter 5

In the case of Fitter Snacker, the sales order clerk must manually check credit. If the clerk fails to do this, then a customer who is a bad risk might receive more credit. In addition, the sales clerk is working with data that are not current, so the credit decision can frequently be made in error. Finally, because it is a manual system, blocked orders may erroneously become real orders, so sales may be affected. With the SAP R/3 system, the check is automatic, the data are up-to-date, and it is a simple matter to review blocked sales orders.

Exercise 5.1

a. Create a document that describes Fitter Snacker's current credit management procedure. This document should be written so that it could be used to train a new employee in the credit management process.

b. Revise the document to reflect the process improvements that would result if Fitter Snacker were performing credit checks using an integrated information system.

PRODUCT PROFITABILITY ANALYSIS

Business managers use accounting data to perform profitability analyses of a company and its products. When data are inaccurate or incomplete, the analyses are flawed. There are three main reasons for inaccurate or incomplete data: inconsistent record keeping, inaccurate inventory costing systems, and problems consolidating data from subsidiaries. In the next section, we'll look at each of these causes.

Inconsistent Record Keeping

Each of FS's marketing divisions maintains its own records and keeps track of sales data differently. When the Direct Division records a sale, the files include a code for a sales region (Northeast, Southeast, and so on). When the Distributor Division records a sale, the files include a code for the state. Suppose that an FS executive asks for a report that summarizes monthly sales dollars for all states in the Mid-Atlantic region for the previous year. Neither division's records are set up to answer that question. An FS accountant would need to perform this tedious series of steps:

1. Go to the source sales documents.
2. Code each document by state and region (as the case may be).
3. Summarize the data by state and region to produce the report. This would have to be done by hand, entering the data into a spreadsheet for review, or by some other means.

Similarly, suppose FS's management wants to evaluate the efficiency of Production's operations. Production uses paper records, so data must be taken from the paper records and entered into a spreadsheet. Many times, those paper records are inaccurate or missing, making the validity of the final report questionable.

There are many variations on this theme. Conceivably, a company's divisions do maintain the same data about a function, but often each division's systems were created at different times, each using a different file system. To answer a manager's question, at least one set of data would need to be rekeyed into a spreadsheet (or some other middleware program) for the merged analysis. Again, it is possible to get an answer, but doing so takes time.

Without integrated information systems, much accounting and management-reporting effort is similar to that in these examples: working around the limitations of the information systems to produce useful output.

With an ERP system, however, this sort of effort would be minimized or eliminated because both divisions would have recorded and stored their data in the same way in the same records. If the company's process was changed to fit the best practices of the software when the system was installed, the managers of each division would have agreed on the way data were stored and collected as part of the system's configuration. Later, questions could be answered in a few minutes by any accountant (or manager or salesperson, for that matter) who understands how to execute a query in the database language or how to use built-in management-reporting tools.

Inaccurate Inventory Costing Systems

Correctly calculating inventory costs is one of the most important and challenging accounting tasks in any manufacturing company. Managers need to know how much it costs to make individual products, so they can identify which products are profitable and which are not.

First, we will provide an overview of inventory cost accounting. Next, we will see how an ERP system can improve the accuracy of inventory cost accounting. Finally, we will discuss the rationale behind activity-based costing as a method to further improve the accuracy of inventory cost accounting.

Inventory Cost-Accounting Background

A manufactured item's cost has three elements: the cost of raw materials, the cost of labor employed directly in the production of the item, and all other costs, which are commonly called **overhead**. Overhead costs include factory utilities, general factory labor (such as custodians or security guards), managers' salaries, storage, insurance, and other manufacturing-related costs.

Materials and labor are often called **direct costs** because the constituent amounts of each in a finished product can be estimated fairly accurately. On the other hand, the overhead items, called **indirect costs**, are difficult to associate with a specific product or a batch of specific products. In other words, the cause-and-effect relationship between an overhead cost (such as the cost of heat and light) and making a particular product (NRG-A bars) is difficult to establish.

Nevertheless, overhead costs are part of making products, so some way to allocate these indirect costs to items made must be found. A common method is to use total machine hours, on the assumption that overhead is incurred to run the machines that make the products. Overhead costs are added up and then divided by the expected total machine hours to get an amount per hour. This value is then used to allocate overhead costs to products. If, for example, FS's overhead per machine hour is $1,000, and 10,000 bars are made in an hour, then each bar made would be allocated 10 cents of overhead ($1,000÷ 10,000). Another allocation method distributes total overhead costs by direct labor hours, on the assumption that overhead costs are incurred so that workers can do their jobs.

Companies such as FS that produce goods for inventory typically record the cost of manufacturing during a period using a standard cost. **Standard costs** for a product are established by studying historical direct and indirect cost patterns in a company and taking into account the effects of current manufacturing changes. At the end of an accounting period, if actual costs differ from standard costs, adjustments to the accounts must be made to show the cost of inventory owned on the balance sheet and cost of inventory sold on the income statement.

For example, FS might determine that each NRG-A bar should cost $0.75 to make—that is, the cost of raw materials, labor, and overhead should equal $0.75, given the budgeted number of units. That amount would be FS's standard cost for a bar. During a month, FS might make 1 million NRG-A bars. Using the standard cost, it would increase its balance sheet inventory account by $750,000. Also, assume that the company sells 800,000 bars in the month. In the income statement, the cost of the sales would be shown as $600,000. Also, the inventory account would be reduced by $600,000, because the company no longer has the units to sell.

If actual costs in the month equal standard costs, no balance sheet or income statement adjustments would be needed. Actual costs never exactly equal expected costs, however, so adjustments are needed. The differences between actual costs and standard costs are called **cost variances**. Note that cost variances arise with both direct and indirect costs. These variances are calculated by comparing actual expenses for material, labor, utilities, rent, and so on with predicted standard costs.

If records for the various elements are kept separately, compiling variance adjustments can be quite tedious and difficult. If products are made by assembling parts that are made at different manufacturing sites, and the sites use different information systems, the adjustments may be very imprecise.

ERP and Inventory Cost Accounting

Many companies with unintegrated accounting systems infrequently analyze cost variances. Often, these companies do not know how much it actually costs to produce a unit of a product. Knowing precisely how much production costs can be very important. Follow this example to explore why.

Suppose FS has an opportunity to sell 300,000 NRG-A bars to a new customer. This is a huge order for FS. The customer wants a price of $0.90 per bar. FS's standard cost per bar is currently $0.75, based on information that is two months old. FS knows that the costs to manufacture snack bars have been increasing significantly in the past months. FS does not want to sell at a loss per unit, but it also does not want to lose a large order or a potentially good customer. Because of the difficulty in compiling all the data to calculate cost variances, FS's cost variances are only analyzed quarterly, and new data will not be available for another month. Should FS accept the large order?

If FS had an ERP system, costs throughout the company would have been recorded in the company-wide database as they occurred. The methods for allocating costs to products and for computing variances would have been built into the system when it was configured. Thus, variances could be computed automatically when needed. Not only would this simplify the process of adjusting accounts, FS's management would always have accurate, up-to-date information on cost variances. FS could decide whether it can profitably sell snack bars for $0.90 each. Furthermore, with a properly operating Sales and

Operations Planning process, FS could determine whether it has the capacity to complete the order on time as well! If overtime is required to complete the order, then the planning capabilities of the ERP system could be used to evaluate costs using overtime production.

ERP system configurations allow costs to be tracked using many bases—by job, by work area, or by production activity. This means that unit costs can be computed using different overhead allocation bases, allowing the analyst to play "what if" with product profitability decisions. In an unintegrated system, doing such multifaceted tracking would be time consuming and difficult.

Activity-Based Costing and ERP

A trend in inventory cost accounting is toward **activity-based costing (ABC)**. In ABC, activities associated with overhead cost generation are identified and then records are kept on the costs *and* on the activities. The activities are viewed as causes (drivers) of the overhead costs. This view treats overhead costs as more direct than traditional cost-accounting methods have treated them. ABC tries to avoid rough allocation procedures in an attempt to assign costs more precisely to individual products. A company using ABC to provide more accurate cost allocations would be able to determine which products have the highest profit margin, information that is crucial in making strategic decisions on product lines. ABC is often used when competition is stiff, overhead costs are high, and products are diverse.

Consider this example from FS's operations. Suppose that storage of raw materials is considered an activity. Assume that storage activities differ between NRG-A and NRG-B bars because the ingredients are different, and that some of these storage activities are more labor-intensive than others. In an ERP system, FS would keep track of the various activities (how often they occur) and the cost of each. Later, when determining the profitability of each kind of bar, storage costs would be added in, based on their cause. This costing is more precise than computing an average storage cost based on total storage costs and machine hours, and then allocating that amount to each kind of bar. Conceivably, if the activities differ enough from one bar to the next, one could be significantly more or less profitable than the other. This fact would be revealed by the ABC approach, but *not* by traditional cost-accounting approaches. Letting managers see that difference is the value of an information system that supports ABC.

Not all overhead costs can be linked to products by their activities. However, many can, depending on the company and the manufacturing situation. For many companies, the cost and effort required to implement ABC is justified by the value of the improved information yielded.

ABC requires more bookkeeping than traditional costing methods because a company must do ABC in addition to traditional costing, and because ABC requires keeping track of instances of activities, not just the costs. Companies often use ABC for strategic purposes, and at the same time use traditional costing for generally accepted accounting principles such as bookkeeping and taxes. Having an integrated information system allows a company to do both kinds of accounting much more easily. A recent study of companies with and without ERP revealed that: (1) ERP companies had nearly twice as many cost-allocation bases to use in management decision-making, and (2) the ERP companies' managers rated their cost-accounting system much higher. Companies with ERP systems value their cost-accounting systems more than if they didn't have ERP, and these companies also have more faith in their numbers from the ERP system.[1]

Companies with Subsidiaries

Some companies have special operations that make closing their books at the end of an accounting period a challenge. Companies that have subsidiaries or branches face such a challenge. Because company managers want the big picture of overall operations and profitability, account balances for each entity must be compiled and forwarded to the home office. A consolidated statement for the company as a whole must then be created.

You might think this would be merely an arithmetic problem: add up cash for all the entities, accounts receivables for all the entities, and so on through the accounts. The job, however, is more difficult than that. Problems can arise from two sources: accounts stated in another country's currency must be converted to U.S. dollars (in the case of a U.S. parent company), and transactions between companies and their subsidiaries must be eliminated from the accounts.

Currency Translation

The following scenario illustrates the problems of **currency translation**. Assume one euro is worth $1.25. A company's European subsidiary reports cash of 1 million euros at the end of the year. When the European subsidiary's balances are consolidated with those of the U.S. parent company at the end of the year, $1,250,000 will be recorded. The same sort of translation would be done for all the European subsidiary's accounts.

A complicating factor is that exchange rates fluctuate daily. An ERP system can be configured to access daily exchange rates and translate daily transactions automatically.

Intercompany Transactions

Transactions that occur between companies and their subsidiaries, known as **intercompany transactions**, must be eliminated from the books of the parent company because the transaction does not represent any transfer of funds into or out of the company.

Suppose that Acme Inc. owns Bennett Manufacturing. Bennett sells raw materials to Acme for $1 million. Acme then uses the materials to make its product. Bennett's sale is Acme's cost of sales. From the point of view of an outsider, money has merely passed from one part of the consolidated company to another. A company cannot make a profit by selling to itself.

Companies often do business with their subsidiaries. If a company does so, then such transactions will occur frequently. Keeping track of them and making the adjustments can be a challenge for the accountants.

Microsoft adopted SAP R/3 worldwide to handle the consolidation of its 130 subsidiaries' accounting data. According to Microsoft, prior to installing SAP R/3, each entity did its own accounting on-site in separate systems, and then transmitted the files for reloading into another system, followed by additional manipulation. It took a significant number of people more than a week to consolidate all the numbers. With SAP R/3, a single person now does the task in a matter of hours. The company had disparate systems all over the world being used for the same business functions, and the consolidation of that information was extremely costly and time consuming. The information came from many different systems, with different field sizes, types of characters, ways of storing information, terminology, and account code block structures. Microsoft had to maintain all sorts of tables at its corporate headquarters to map subsidiary systems to the appropriate corporate account structures, consolidate the information, and then reconcile any differences. This had a severe impact on getting Microsoft's books closed and in providing management with timely numbers. Microsoft adopted a single version of SAP R/3 for all its companies to ensure consistency of business processes worldwide. Now all of Microsoft's subsidiaries are using SAP R/3 to report their financial activities directly. At any time, Microsoft can look directly at financial activity around the world. Microsoft reports that the cost of the ERP implementation will be recovered in two years through reduced costs.[2]

MANAGEMENT REPORTING WITH ERP SYSTEMS

The integrated nature of an ERP system and the use of a common database and built-in management reporting creates numerous benefits. Although reporting accounting information seems a trivial task, it is often very challenging for companies to generate the right reports for the right situation. Without an ERP system, the job of tracking all the numbers that need to go into a report is a monumental undertaking. With an ERP system, a vast amount of information is available for reporting purposes. Often companies take years after ERP implementation to figure out which reports are the most critical for decision making. In this section, we will examine some management reporting and analysis tools available with ERP systems.

Using Document Flow for Customer Service

As you have seen, when an ERP system is used, all transactions in all areas of a company get posted in a centralized database. It is worth reemphasizing that the database *is* the company's "books." There is no separate set of books for Marketing or Production or Purchasing.

Thus, even though it is common usage to refer to "data flows" in an ERP system, it is actually a misnomer. Data do not flow from one ERP module to another because they are all in one place—the database. Each area views the same records. It might be better to speak of "data access" rather than "data flows" when talking about how these areas use the common database. However, typical usage by companies (*data flows* and *data sharing*), is probably too ingrained to avoid.

As you learned earlier, each transaction that is posted in SAP R/3 gets its own unique document number. This number allows quick access to the data. If you need to look up a transaction online, you do so by referencing the document number, which acts as an index into the appropriate database table entries.

In SAP R/3, document numbers for related transactions are associated in the database. This provides an electronic audit trail for analysts trying to determine the status of an order. The best example of this concept is the linkage of document numbers for a sales order. Figure 5-6 illustrates the document flow concept.

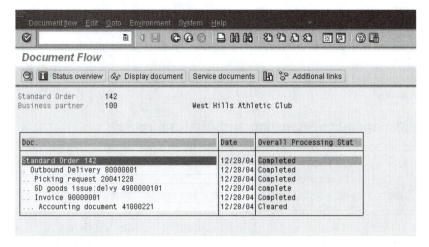

FIGURE 5-6 Document flow of a transaction in SAP R/3

The linked events shown in Figure 5-6 progress as follows:

1. When the order was placed, sales order document 142 was created.
2. The system recorded the Delivery, which is the transfer of the order's requirements to the Materials Management module. It is denoted by document 8000001.
3. The picking request, which is the document that tells warehouse personnel which items make up the order, was created on Dec. 28, 2004, and given document number 20041228.
4. The goods were removed from inventory on the same day, an event recorded by document number 4900000101.
5. After the goods were issued, an invoice was generated so that the customer would be billed. The invoice was given document number 90000001.
6. At the same time, the accounting entries for the sale were generated. The posting document number is 41000221.

The document flow can be used to drill down to see the details of any one of these events. The term **drill down** refers to the capability of viewing summary information and then being able to look at details behind the summary. For example, the user can double-click the order number (142) and see the details of the order—products ordered, quantity, customer name, and so on. From that display, the user can double-click the product numbers or the customer number to see details about them. To see the debits and credits in the accounting entry, you could double-click 41000221 to see the scheduled entries.

Users can access the document flow from any SAP screen. If a customer were to call and ask about the status of an order, the clerk could access the document flow and see whether the goods had been shipped. Similarly, a customer could call with an invoice number and questions about the billing. The customer service representative could use the document flow to access documents in the chain of events, such as the original order or the picking request. This sort of research can be done quickly with SAP R/3. With unintegrated systems, establishing the audit trail and researching source documents can be very difficult and time consuming.

Using Built-In Management-Reporting and Analysis Tools

Accounting records are maintained in the common database. The advantage of using a database is the ability to query the records to produce standard reports as well as answer ad hoc questions. An ad hoc question is one that is spontaneous. For example, a Fitter Snacker manager might run into an analyst's office and ask for an immediate sales report for the third quarter, snack bar division, by product. Traditional accounting packages are not optimized to set up and execute queries against accounting records, but database packages are. Therefore, when the records are kept in a database, the user gets a double benefit. The records can be kept in an accounting package *and* the records can be queried because of the built-in database language.

Thus, if a user wants to identify the 10 largest orders placed by Health Express in the past year, a query could be executed to show the answer. In principle, this query could directly access the transaction records to get the answer. In practice, this would mean that analysts running queries would be accessing the records at the same time as current transactions are being recorded. This competition can slow down processing in even a large database, such as those used by ERP packages.

SAP's solution to this problem is to provide a **data warehouse** within each major module. Analysts query the warehouse rather than the transaction database. For example, SAP R/3 provides the Sales Information System (SIS) tool for analysts querying the sales records and the Logistics Information System (LIS) tool for analysts querying the logistics (shipping) records. Both the SIS and the LIS come embedded with SAP R/3. Also, as mentioned in Chapter 2, SAP also sells its Business Warehouse (BW) product, a completely separate information system that extracts data from the SAP R/3 system. With BW, users have great flexibility in defining data reports and analyses in a system that does not compete for system resources with transaction processes.

THE ENRON COLLAPSE

On Oct. 16, 2001, Enron, then one of the world's largest electricity and natural gas traders, reported a $618 million third-quarter loss and disclosed a $1.2 billion reduction in shareholder equity, partly related to partnerships run by its Chief Financial Officer (CFO) Andrew Fastow. Until that time, Enron had been a rapidly growing firm that was revolutionizing the energy business and making millionaires out of its investors. CEO Jeffrey Skilling, who resigned on Aug. 14, 2001 for personal reasons, had helped transform the company from a natural gas pipeline company to a global marketer and trader of energy. Employees who worked for the company had been encouraged to invest large portions of their 401K retirement savings accounts in Enron stock by matching employee contributions. On Oct. 17, the day after Enron reported its tremendous third-quarter loss, the U.S. Securities and Exchange Commission (SEC) sent a letter to Enron asking for information about the loss. The SEC is the U.S. Securities and Exchange Commission, dedicated to protecting investors and maintaining the integrity of the securities markets.[3] Enron's high-flying business practices immediately began unraveling.

On Oct. 22, 2001, Enron acknowledged an SEC inquiry into a possible conflict of interest related to the company's dealings with the partnerships run by CFO Fastow. Shares of Enron sank more than 20 percent on the news. Two days later, Enron ousted CFO Fastow. On Nov. 8, Arthur Andersen, Enron's auditing firm, received a federal subpoena for documents related to Enron, and on Dec. 2, Enron made the largest Chapter 11 bankruptcy protection filing in history. Clearly, the accounting records made public from Enron did not reflect the financial health of the company.

Enron began as an oil pipeline company in Houston in 1985. With the deregulation of electrical power markets, Enron expanded into an energy broker, trading electricity and other commodities. Unlike a traditional exchange that brings together buyers and sellers, Enron entered into separate contracts with sellers and buyers, making money on the difference between the selling price and the buying price. Enron kept its books private, making it the only party that knew both prices. Enron's business developed over time, allowing firms to insure themselves against risk due to a range of factors, including changes in interest rates, weather, or a customer's inability to pay. The volume of these financial contracts was far greater than the volume of contracts to actually deliver commodities.

To manage the risk in these contracts, Enron was employing Ph.D.s in mathematics, physics, and economics. Risk management balances opportunities in business against risks taking that business. As Enron's services became more complex and its stock soared, Fastow created partnerships between Enron and companies involved in Internet broadband technologies, computer technology, and energy, to name a few. Some of the partnerships were faked to mask billions of dollars in debt, allowing managers to shift debt off the books.

The partnerships that Fastow engineered were the subject of discussion well before Enron's bankruptcy. In a June 1, 1999 article in *CFO* Magazine, Ronald Fink noted that the Financial Accounting Standards Board was looking at rule changes that would affect companies using creative financing techniques like Enron. Enron owned a number of subsidiaries, but made sure that it owned no more than 50 percent of the voting stock. As a result, Enron was able to keep the debt and assets of these subsidiaries off Enron's own books. If Enron had not been able to use these creative accounting practices, the company would have had to report a much higher percentage of debt than it reported by keeping the subsidiaries' debts off its books, which would have increased the costs that Enron paid to borrow money.

Enron's financial statements for years had been audited by Arthur Andersen, a highly-regarded accounting firm. As Enron's auditor, Andersen issued annual reports attesting to the validity of Enron's financial statements and was supposed to function as an unbiased, incorruptible observer and reporter. Enron's Oct. 15, 2001 press release characterized numerous charges against income for the third quarter as "non-recurring" even though Andersen believed the company did not have a basis for concluding that the charges would in fact be non-recurring. Indeed, Andersen advised Enron against using that term, and documented its objections internally in the event of litigation, but did not report its objections or otherwise take steps to cure the public statement.

Perhaps the most damning part of Andersen's indictment was the destruction of documents. On Oct. 22, 2001, Enron acknowledged a Securities and Exchange Commission inquiry, and on Oct. 23, Andersen personnel were called to urgent and mandatory meetings. Andersen employees on the Enron engagement team were instructed by Andersen partners and others to immediately destroy documentation relating to Enron, and were told to work overtime if necessary to accomplish the destruction. During the next few weeks, an unparalleled initiative led to the shredding of paper documentation and the deletion of hundreds of computer files.

Outcome of the Enron Scandal

The effect of the Enron scandal goes both deep within and well beyond the company. Most of Enron's more than 20,000 creditors will receive about one-fifth of the approximate $63 billion they are owed. Shareholders will get nothing, which means that all employees that invested in the company's 401K program have lost their investment. Thousands of workers lost their jobs, and 31 individuals are either awaiting trial on or have pleaded guilty to criminal charges. Those awaiting trial include Enron founder and former chairman Kenneth Lay and former CEO Jeffrey Skilling. Both have pleaded innocent to charges that include fraud and conspiracy. J. Clifford Baxter resigned as Enron vice chairman on May 2, 2001. He was found shot to death in a car Jan. 15, 2002, in an apparent suicide. Andrew Fastow faces dozens of charges, including fraud, money laundering, and conspiracy. On June 15, 2002, jurors convicted Arthur Andersen for obstructing justice when it destroyed Enron documents while on notice of a federal investigation. Andersen had claimed that the documents were destroyed as part of its housekeeping duties and not as a ruse to keep Enron documents away from the regulators. That October, U.S. District Judge Melinda Harmon sentenced Andersen to the maximum: a $500,000 fine and five years' probation. Those events were anticlimactic, however, as the former auditing giant was all but dismantled by then. Once a world-class firm with 28,000 employees in the United States alone, Andersen has since been whittled down to about 250 people, most of them dealing with litigation and running a training center outside Chicago.

As a result of the failure of Enron, as well as the high-profile bankruptcies of WorldCom and Global Crossing, the U.S. Congress passed the Sarbanes-Oxley Act of 2002. This act was designed to prevent the kind of fraud and abuse that led to the Enron downfall.

Key Features of the Sarbanes-Oxley Act

The Sarbanes-Oxley Act is designed to encourage top management accountability. Frequently, top executives involved in corporate scandals claim that they were unaware of abuses occurring at their company. Title IX of the Sarbanes-Oxley Act adds the requirement that financial statements filed with the Securities and Exchange Commission must include a statement signed by the Chief Executive Officer and Chief Financial Officer that the financial statement complies with the SEC rules. Specifically, the statements certify that "the information contained in the periodic report fairly presents, in all material respects, the financial condition and results of operations of the issuer."[4] The penalty for "willfully certifying any statement...knowing that the periodic report accompanying the statement does not comport with all the requirements set forth in this section shall be fined not more than $5,000,000, or imprisoned not more than 20 years, or both."[5]

Title II of the act addresses auditor independence. Among other things, this section of the act limits the non-audit services that an auditor can provide. Among the non-audit services prohibited by the act are:

- Bookkeeping or other services related to the accounting records or financial statements
- Financial information systems design and implementation services
- Legal services
- Expert services unrelated to the audit
- Management functions
- Human resources functions
- Any other service that the Public Company Accounting Oversight Board (PCAOB) determines to be impermissible. The PCAOB was created in Title I of the act with broad powers to regulate audits and auditors of public companies.

Title IV of the act is Enhanced Financial Disclosures, which specifies more stringent requirements for financial reporting. Section 404 of Title IV requires that a public company's annual report contains management's internal control report. This control report outlines management's responsibility for establishing and maintaining adequate internal control over financial reporting and assesses the effectiveness of its internal control over financial reporting. Section 409 of Title IV addresses the timeliness of reports, and may require companies to file an SEC report within two days of a significant trigger event, for example, completion of an acquisition or default by a major customer.

IMPLICATIONS OF THE SARBANES-OXLEY ACT FOR ERP SYSTEMS

Certainly the Sarbanes-Oxley Act has significant implications for a firm's information systems. To meet the internal control report requirement, a company must first document the controls that are in place and then verify that they are not subject to error or manipulation.

An integrated information system provides the tools to implement internal controls as long as the system is configured and managed correctly. However, even the passage of the Sarbanes-Oxley Act and the availability of state-of-the-art ERP technology cannot prevent insidious and systematic fraud similar to the Enron scandal. An ERP system relies on a central database with accurate information. It would be more difficult to hide fraudulent dealings with an ERP system, and perhaps Enron's problems would have been more obvious to stakeholders of the company. But it is unlikely that an ERP system can prevent all fraud.

On the positive side, companies with ERP systems in place will have an easier time complying with the Sarbanes-Oxley Act than will companies without an ERP. The next section explores ways in which SAP R/3 and other ERP systems can prevent corporate fraud and abuse. Systems from vendors other than SAP have functionality similar to that of SAP R/3.

Archiving

One of the first things a new SAP R/3 user notices is that the software offers very few ways to delete items. For example, the menus in the SAP R/3 system related to material master data (master records that describe material characteristics) are shown in Figure 5-7. There are options for creating, changing, and displaying, but not simply deleting—the closest option is to flag for deletion. Before a material can be deleted from the SAP R/3 system, a user must create an auditable record of its existence. Data are removed from the SAP R/3 system only after they have been recorded to media (tape backup, DVD-R) for permanent storage. This permanent storage, or **archive**, allows auditors to reconstruct the company's financial position at any point in the past.

FIGURE 5-7 Transaction options for material master data

Suppose data could be freely deleted from the system. An unscrupulous employee could create a fictitious vendor, post an invoice from the vendor, have payment made for the bogus invoice to a Swiss bank account, and then delete all records of this transaction. It would be very hard to detect the fraud and probably impossible to find out who committed it, because the records would no longer exist.

Not only does the SAP R/3 system require archiving before data can be deleted, but it also keeps track of when data are created or changed. Figure 5-8 shows the Change Record for the material master. Each time the material master is changed, the Change Record tracks the change that was made, who made the change, and when the change was made.

FIGURE 5-8 Change Record for material master

User Authorizations

Another way that an ERP system can prevent an unscrupulous employee from making payments to a fictitious vendor is through user authorizations and separation of duties. SAP R/3 has sophisticated user administration tools to ensure that employees can perform only the transactions required for their jobs. The SAP R/3 system provides tools for different levels of authorization management. One way that the SAP system controls user authorizations is through the Profile Generator. The Profile Generator provides a simple method for selecting the functions that a user should be allowed to perform.

Figure 5-9 shows a predefined role in the SAP R/3 system for a user whose job involves managing material masters and bills of material. If a user is assigned the task of managing material masters and bills of material, then that person can perform any transactions shown on the role menu in Figure 5-9.

Managing authorized usage can be one of the most complex tasks in an ERP system. The joke in the world of authorizations is "If you're doing your job, we're not doing ours!", meaning that it is frequently a challenge to provide users with the proper authorizations in a timely manner. Most companies, however, take the approach that it is better to err on the side of taking a little longer to grant the right level of authorization rather than giving users too much authority quickly.

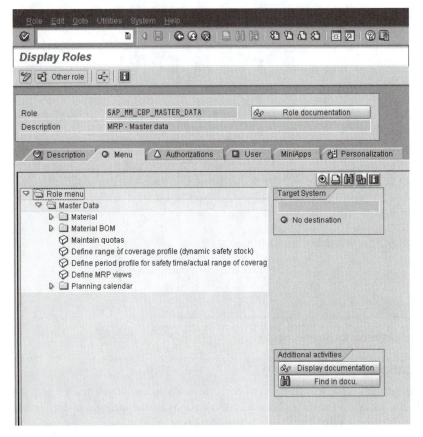

FIGURE 5-9   Role for material management master data

Tolerance Groups

Another ERP way to make sure employees do not exceed their authority in financial transactions is by setting limits on the size of transaction an employee can process. In the SAP R/3 system, this is done using tolerance groups, as shown in Figure 5-10. As you learned in Chapter 2, tolerance groups are preset limits on an employee's ability to post transactions. Tolerance groups set limits on the dollar value for a single item in a document as well as the total value of the document. Just as importantly, it sets a limit on payment differences. Suppose a customer has been invoiced for $1,005 but accidentally sends in a check for $1,000 to pay the invoice. The cost of requesting and processing a second payment for the $5.00 would cost both parties more than $5.00. In this case it is better to accept the $1,000 check as payment in full and account for the difference as a variance. In Figure 5-10, the Permitted payment differences section shows that the system would allow the user to process a payment that was in error by no more than 1 percent or $10.00. Notice that the group field is blank in Figure 5-10. That means that this group is the default tolerance group. If an employee is not assigned to any other tolerance group, then by default the limits in the default group apply. As with authorizations, it is a safe policy to define a default tolerance group with low limits and err on the side of less authority rather than more authority.

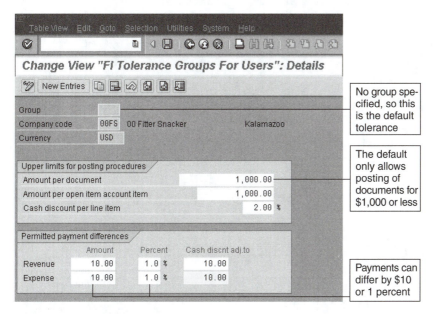

FIGURE 5-10   Default tolerance group

Financial Transparency

A key feature of any ERP system is the ability to drill down from a report to the source documents (transactions) that created it. For example, if sales figures for a region look unusually high, the user can double-click the figure in the report and drill down to review the specific sales orders that comprise the overall sales figures to verify the results. The ability to drill down from reports to transactions makes it easier for auditors to confirm the integrity of the reports. Figure 5-11 shows a general ledger account balance for raw material consumption. This general ledger accounts for all raw material usage at the company.

To see where the figures in this report come from, a user can double-click the line for Period 9, which takes you to the screen in Figure 5-12.

This figure shows that two items make up the $8,810.00 raw material consumption. A manager might be intrigued by the $10 expense and want more information on what caused the expense. Clicking the Detail button provides more detailed information on the $10 expense, as shown in Figure 5-13. Fifty pounds of material were charged to cost center R010, which is a research and development cost center. From this screen even more details can be displayed.

With a few mouse clicks, an auditor can move from a summary statement of a general ledger account to find out all of the details related to an employee in the research and development department taking 50 pounds of oats from inventory for product development. With a properly configured and managed ERP system, there are direct links between the company's financial statements and the individual transactions that make up the statements, so that fraud and abuse can be detected more easily.

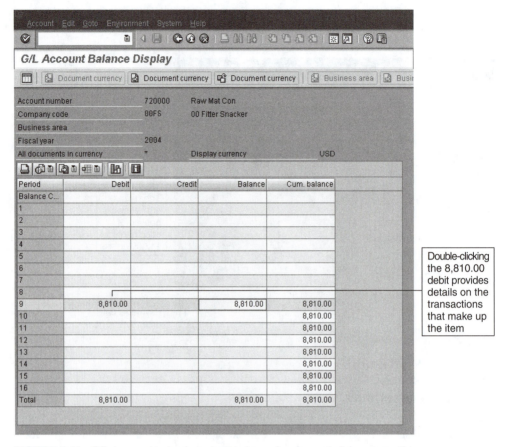

FIGURE 5-11 G/L (general ledger) account balance for raw material consumption

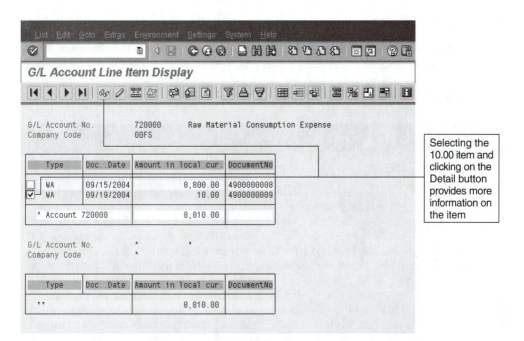

Selecting the 10.00 item and clicking on the Detail button provides more information on the item

FIGURE 5-12   Documents that make up G/L account balance for raw material consumption

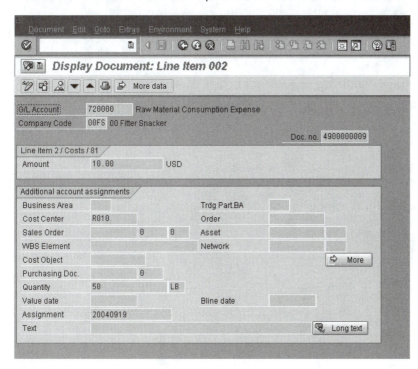

FIGURE 5-13   Details on $10.00 line item in G/L account for raw material consumption

Exercise 5.2

Assume that you are the manager of the Accounting department at Fitter Snacker. What changes must you make in your accounting practices to prepare for the Sarbanes-Oxley Act? Write a proposal to your manager outlining your plan to be ready for the additional amount of reporting necessary.

ANOTHER LOOK

The One-Day Close

As you learned in the beginning of this chapter, one important function of the Accounting department is closing the books. Having integrated, accurate information at your fingertips greatly simplifies closing the books. Some companies strive to be able to close their books in one day; for most companies, it takes days or weeks or even months to get all the figures to balance. Some even perform a virtual close (simulating the close on the computer) at various times during the month, to see how well they are doing financially. How do companies do this?

One company able to close the books in less than one day is Cisco, the manufacturer of networking hardware. Five years ago, it took Cisco accountants an average of two weeks to close the books. Without a single integrated system, accountants were chasing numbers from various systems around the company. The information was difficult to track down and was often inconsistent between systems. All companies find that inaccuracies in financial statements can happen more often with unintegrated systems.

With the implementation of its Oracle ERP system, the one-day close is a reality for Cisco. As an added benefit, the time devoted to transaction processing has fallen to 35 percent from 65 percent and finance group expenses as a percentage of total company revenue have fallen to 1.3 percent from 2 percent.[6]

The potential for further streamlining exists. Todd M. Barnhart of Darwin Publications notes that while companies are already gaining great benefits from streamlining their supply chains with ERP, it is now time to streamline their financial supply chains. Barnhart argues that if companies have excellent information on their finances, they can hold less cash, in the same way that accurate information on their supply chain allows them to hold less inventory.[7]

Question:

1. Research the concept of the one-day close. Find out what types of companies can achieve this, and report to your instructor on how they can manage their books so quickly and accurately.

Chapter Summary

1. Companies need accounting systems to record transactions and generate financial statements. The accounting system should let the user summarize data in meaningful ways. The data can then be used to assist managers in their day-to-day work and in long-range planning.

2. With unintegrated information systems, accounting data might not be current, and this can cause problems when trying to make operational decisions, such as granting credit. Data can also be inaccurate because of weaknesses in unintegrated systems, and this problem can have an effect on decision-making and therefore on profitability.

3. Closing the books at the end of an accounting period can be difficult with an unintegrated IS, but it is relatively easy with an integrated IS. Closing the books means zeroing out the temporary accounts.

4. Using an integrated IS and a common database to record accounting data has important inventory cost-accounting benefits. More precise record keeping is possible, and this can lead to more accurate product cost calculations. These, in turn, can help managers determine which products are profitable and which are not.

5. The use of an integrated system and a common database to record accounting data has important management-reporting benefits. The user has built-in drill-down and query tools available as a result.

6. The introduction of the Sarbanes-Oxley Act, a 2002 U.S. federal regulation written and passed in the wake of the Enron collapse, promotes management accountability by requiring extra financial approval and reporting. Because ERP systems can help companies meet the requirements of this legislation, the act has increased the need for integrated data reporting.

Key Terms

Activity-based costing (ABC)	Financial Accounting
Archive	General ledger
Balance sheet	Income statement
Cost variance	Indirect costs
Currency translation	Intercompany transaction
Data warehouse	Managerial Accounting
Direct costs	Overhead
Drill down	Standard cost

Exercises

1. Describe the differences between Financial and Managerial Accounting. How can ERP systems benefit workers in both?

2. This exercise tests your understanding of the information needed to trace a sale through a multistep ERP sales cycle: sales order, inventory sourcing, delivery, billing, and payment.

Assume that an order has been placed with your company's ERP system. These events occur:

a. The customer's credit is automatically checked and found to be acceptable. The order is recorded for the delivery date requested.

b. The system schedules the production of the goods. (There is not enough inventory to ship from stock.)

c. The system schedules raw material orders from the vendors to make the goods ordered.

d. The raw materials are received and stored.

e. The goods are produced and reserved for shipment to the customer.

f. The delivery is scheduled, and the goods are put on the delivery truck. An invoice is printed and included with the shipment.

g. Accounting is notified of the shipment's details.

h. A month later, the customer sends in payment, which is recorded in accounting.

For each of these events, list the information that must be recorded in the common database. You do not need to know how to use a database to do this, nor do you need to understand bookkeeping. At each step, did the wealth of the company increase or decrease? At each step, how did the company's obligations to outsiders change? At each step, how did the obligations of outsiders to the company change?

3. Suppose it is 4 p.m., September 29, 2005. Fitter Snacker's CEO sends this e-mail message to the accounting manager: "I need to meet with the Board of Directors tomorrow morning. The Board members are concerned about the current sales of our energy bars. They would like to see sales data for today, 9/29, as a typical day. Please complete the attached report and have it on my desk by 9 a.m. tomorrow, September 30." The blank table shown in Figure 5-14 is attached to the memo.

Sales data September 29	Distributor Division	Direct Sales Division	Total
# Bars sold			
NRG-A			
NRG-B			
Total			
$ Value of bars sold			
NRG-A			
NRG-B			
Total			
# Customers sold to			
NRG-A			
NRG-B			
Total			

FIGURE 5-14

Given the company's sales-order data-processing practices, why won't the accountants be able to answer this question easily? Cite the practices that cause difficulty and explain. If FS had an ERP system, why would this question be easy to answer? To answer this question, review Chapters 3 and 4 to see how sales orders are processed.

4. The following exercise will test your understanding of FS's current credit-check system. In each situation, you are given background data and information about documents in the system.

SITUATION 1

Background data	
Today's date	6/29/05
Current list price, NRG-A	$1.00/bar
Current list price, NRG-B	$1.10/bar
Accounts receivable balance at start of business day, ABC Corp.	$8,000
Credit limit, ABC Corp.	$10,000
Current order	
Product	NRG-A
Amount	4 cases (1,152 bars)
Price	List
Ship to	ABC headquarters
Date desired	7/5/05
Next invoice number	A1001
Documents in system	

No documents relating to ABC are in the system.

a. Given the state of the system, will credit be granted or denied for ABC Corporation's current order?

b. If the system processed data in a more timely way, should credit be granted or denied?

SITUATION 2

Background data	
Today's date	7/3/05
Current list price, NRG-A	$1.00/bar
Current list price, NRG-B	$1.10/bar

Background data

Accounts receivable balance at start of business day, KLM Corp.	$5,000
Credit limit, KLM Corp.	$7,000

Current order

Product	NRG-B
Amount	5 cases (1,440 bars)
Price	List
Ship to	KLM headquarters
Date desired	7/8/05
Next invoice number	A1200

Documents in system

Purchase order KLM 82332 for three cases (864 bars) of NRG-A. This order is in the sales order-entry program, but it has not been transferred to the accounting program (thus, Accounting does not yet know about this sale).

c. Given the state of the system, will credit be granted or denied for KLM Corporation's current order?

d. If the system processed data in a more timely way, should credit be granted or denied?

SITUATION 3

Background data

Today's date	7/13/05
Current list price, NRG-A	$1.00/bar
Current list price, NRG-B	$1.10/bar
Accounts receivable balance at start of business day, ACORN Corp.	$5,000
Credit limit, ACORN Corp.	$5,000

Current order

Product	NRG-A
Amount	150 boxes (3,600 bars)
Price	List
Ship to	ACORN headquarters

Current order	
Date desired	7/15/05
Next invoice number	A1300

Documents in system

A check from ACORN was received in yesterday's mail and entered into the accounting system. The check is for $2,000, applied to invoices from June's sales. The sales clerks are working from credit-limit printouts prepared at the beginning of the week (two days ago).

    e.   Given the state of the system, will credit be granted or denied for ACORN Corporation's current order?

    f.   If the system processed data in a more timely way, should credit be granted or denied?

5.   ERP systems save time for accountants in many ways. Some researchers have thought that employment of accountants would diminish with the implementation of an ERP system. In reality, accountants are needed more than ever in industry. Explain how the Sarbanes-Oxley Act has impacted the demand for accountants. Research the job market for accountants to prove this point.

6.   Companies usually prepare division and company-wide budgets each year. These budgets show important monthly results that the company plans to achieve: sales, cost of sales, inventory levels, cash on hand, and other key data. Such budgets are effective for planning and controlling operations if they are designed as "flexible" budgets. With a flexible budget, the budget is restated as conditions change from month to month, so goals can be reasonably achieved and used for evaluating performance. If information systems are unintegrated, getting data from departments within the company to create the budget is a chore. Keeping the data current is very difficult and often is not done. If not kept up to date, budgets are not useful for planning and controlling operations. Thus, the "flexible budget" concept is a good idea, but it is difficult to achieve. ERP makes flexible budgeting more achievable, and so ERP helps management discharge its planning and controlling roles better. Why do you think flexible budgeting would be more achievable with ERP systems? List your reasons and explain.

7.   Review FS's unintegrated production and purchasing procedures, described in Chapter 4. How would its current job-scheduling, production, and purchasing procedures result in variances from standard costs? Why would FS have trouble researching these costs at month's end to adjust the "standard costs per unit" to accurate "actual costs per unit"?

End Notes

[1] Kip R. Krumweide and Win G. Jordan, "Reaping the Promise of Enterprise Resource Systems," *Strategic Finance* (October 2000): 49–52.

[2] "Success Story: Microsoft," online and previously archived at *http://www.erpworld.org/erp4a.html*.

[3] http://www.sec.gov/about/whatwedo.shtml.

4   "Sarbanes-Oxley Act, H.R. 3763, Title III, Section 302, §1350 (a)(3).

5   "Sarbanes-Oxley Act, H.R. 3763, Title IX, Section 906, §1350 (c).

6   John S. McClenahen, "The Book On The One-Day Close," IndustryWeek.com (April 1, 2002).

7   Todd M. Barnhart, *The Financial Supply Chain*, Darwin Publications (April 2004).

Additional Resources

Background on the Enron collapse: reporting by *Washington Post*, compiled by Robert O'Harrow, Jr.; research by Lucy Shackelford.

Ronald Fink, "Balancing Act: Will a new accounting rule aimed at off-balance-sheet financing trip up Enron?" *CFO* magazine (July 1, 1999).

Mary Flood, "Andersen conviction affirmed: Appeals court upholds verdict in Enron case," *Houston Chronicle*, (July 1, 2004).

John Goff, "They Might be Giants: It's been nearly two years since Arthur Andersen went under and Sarbanes-Oxley was passed. Have the Big Four audit firms changed since then?" *CFO* magazine (January 12, 2004).

KPMG LLP, "Sarbanes-Oxley: A Closer Look" (January 2003).

KPMG LLP, "Sarbanes-Oxley Section 404: Management Assessment of Internal Control and the Proposed Auditing Standards" (March 2003).

"Timeline of Enron's Collapse," WashingtonPost.com (July 9, 2004).

CHAPTER **6**

HUMAN RESOURCES
PROCESSES WITH ERP

LEARNING OBJECTIVES

After completing this chapter, you should be able to:

- Explain why the Human Resources function is critical to the success of a company.
- Describe the key processes managed by a Human Resources Department.
- Describe how an integrated information system can support effective Human Resources processes.

INTRODUCTION

A company's employees are its most important resources. The Human Resources (HR) Department is responsible for many of the activities that a company performs to attract, hire, reward, and, occasionally, terminate employees. The decisions made in the HR Department can affect every department in the company. Companies are increasingly aware of the importance of an experienced, well-trained workforce and have begun using the term **human capital management (HCM)** to describe the tasks associated with managing a company's workforce.

As a company grows from a small business to a large organization, the need for an organized and effective HR department becomes extremely important. The responsibilities of an HR department usually include:

- Attracting, selecting, and hiring new employees using information from resumes, references, and the interview process.

- Communicating information regarding new positions and hires throughout the organization and beyond.

- Ensuring that employees have the proper education, training, and certification to successfully complete their duties.

- Handling issues related to employee conduct.

- Making sure employees understand the responsibilities of their jobs.

- Using an effective process to review employee performance and determine salary increases and bonuses.

- Managing the salary and benefits provided to each employee and confirming that the proper benefits are disbursed to new and current employees.

- Communicating changes in salaries, benefits, or policies to employees.

- Supporting management plans for changes in the organization (expansion, retirements and so on) so that competent employees are available to support business processes.

Making sure that these tasks are accomplished and valid information is communicated requires an effective information system to control the flow of information. In this chapter we will explore the role of an integrated information system in human resources.

PROBLEMS WITH FITTER SNACKER'S HUMAN RESOURCES PROCESSES

As with Fitter Snacker's other processes, personnel management relies on paper records and a manual filing system. This creates problems because information is not readily accessible or easily analyzed. The HR department's recruiting, hiring, and post-placement processes would operate more efficiently with an integrated system.

Recruiting Process

Fitter Snacker has three employees in its HR department. Problems occur because of the large number of HR processes (from hiring and firing to managing health benefits), the disconnected departments, and the number of people with whom HR interacts. Many of these problems are the result of inaccurate, out-of-date, and inconsistent information.

For example, suppose a department has an opening for a new employee. The department supervisor communicates this need to the Human Resources department by filling out a paper job vacancy form that describes the position and lists the qualifications a candidate must have to fill the position, the type of position (temporary, part-time, full-time, or a co-op/intern), and when the position will become available. Human Resources takes this information, verifies that the position needs to be filled, and gets final approval from the president of Fitter Snacker to begin the recruiting process. Because there is no central information system, the details on the job vacancy form are frequently inconsistent among, and sometimes within, departments.

Usually the job is first posted internally, so that current employees have the opportunity to apply for the position. If no current employees are acceptable for the position, then Fitter Snacker posts the position externally.

A number of problems can arise in the recruiting process. First, the description of the qualifications required for the job may be incomplete or inaccurate, sometimes because the supervisor is in a hurry, sometimes because the supervisor does not realize all of the functions required for the position, and sometimes because the supervisor assumes that all candidates will have certain basic skills. Second, if the job vacancy form is lost or not routed properly, the Human Resources department will not know that the position is available, while the supervisor assumes that the paperwork is in process. The department ends up shorthanded, creating tension or animosity between the departments. Obviously this problem will occur more frequently when job openings are circulated by paper. With an integrated information system, the job information is available immediately and more easily monitored. Another serious recruiting problem related to a paper-based hiring process is the potential loss of a good candidate due to drawn-out hiring practices or lost data.

While Fitter Snacker does not use recruiting agencies or Internet job sites such as Monster.com to find candidates, it does use several other methods to find people for its jobs. FS publishes its job vacancies on the company's Web site, in local newspapers, and, in the case of a professional position, in national publications. In addition, a representative from the HR department attends career fairs and recruits on college campuses for prospective candidates. Occasionally referrals are made by other Fitter Snacker employees, and sometimes individuals searching for open positions at Fitter Snacker send their resumes unsolicited.

Keeping track of resumes and applications is a challenging task. Fitter Snacker has dozens of jobs with different titles and descriptions. The HR department must classify and file all applications and resumes according to the appropriate description. For example, if the resume for a mechanical engineer applying for a job at Fitter Snacker is accidentally filed with resumes of candidates applying for jobs in the accounting department, the mistake may not be discovered in time to include the engineer in the search process. In this case, Fitter Snacker may not hire the best person for the position and the mistake might damage Fitter Snacker's reputation. Fitter Snacker receives dozens of resumes and applications each day, and filing these resumes is a continuing challenge.

Keeping the applicant's data on a paper form means that retrieving the applicant data and using it to evaluate candidates is also a challenging task. To generate a list of potential candidates, Human Resources evaluates resumes and applications it receives in response to a job posting, plus it reviews filed applications that are less than one year old. These resumes and applications must be photocopied and then circulated through the department making the job request. Frequently, more than one person in the requesting department reviews the applicants, and because the applicant data is on paper, the applicant files are reviewed sequentially, slowing the review process.

The Interviewing and Hiring Process

At Fitter Snacker, the requesting department develops a **short list** of candidates for the position by selecting up to three applicants based on the data provided by HR. Human Resources contacts the candidates on the short list, schedules interviews, and creates a file for each candidate. A candidate's file includes a form that shows when the application was received, the position(s) applied for by the candidate, and the date and time of any interviews. If this is the second time the candidate has applied for a job with the company, the form indicates the current status of the candidate: whether the candidate was interviewed and rejected, whether the candidate rejected a job offer, and so on.

If a candidate accepts the interview offer, the HR department makes the arrangements for the job candidate, including travel arrangements and a schedule of interview activities. A representative from the HR department conducts an interview that includes a discussion of the applicant's experience and questions relevant to the position for which the candidate has applied. In addition to the interview conducted by the Human Resources representative, the supervisor of the department in which the position exists also interviews the candidate. Other employees in the department are usually given time to talk to the candidate as well. For most professional positions at Fitter Snacker, the candidate interviews with the plant manager and, frequently, the company president.

After the initial interview process, HR updates the candidate's file to indicate whether the candidate is still a possibility for hire. In some cases, a second interview is scheduled. Once HR has interviewed all the candidates on the short list, an HR person and the supervisor of the requesting department decide which candidates on the short list are acceptable and rank them. If there is an acceptable candidate, the HR person makes the highest-ranking candidate a verbal job offer over the phone. If the candidate accepts the verbal offer, then a written offer letter is sent, which the candidate must sign and return. Once the candidate formally accepts the written offer, his or her file is again updated showing that the candidate has accepted the offer and will begin employment with the company on a specified date.

If there are no acceptable candidates from the short list, or if none of them accept the job offer, then the process must be repeated, which at a minimum would require the development of a new short list but may involve starting over with a new job posting.

Many of Fitter Snacker's problems in the interviewing and hiring process have to do with information flow and communication. Fitter Snacker does not have group appointment calendar software, which would allow HR staff to easily find a time when all key personnel would be available to interview a candidate. Scheduling interviews is frequently a cumbersome process, as the Human Resources employee has to coordinate the interview schedule between the candidate and the appropriate personnel at Fitter Snacker. Because this is done by e-mail and phone, it can take days and sometimes weeks to schedule an interview. A similar problem occurs after the interviews have been completed. Gathering feedback from all involved parties and ranking the candidates takes time and may require multiple meetings. Managing the travel arrangements and reimbursing candidates for their travel expenses are also cumbersome tasks. More than once, a promising candidate has accepted a position with another company because of delays in the Fitter Snacker interviewing and hiring process.

After the candidate accepts the formal job offer, Fitter Snacker hires an HR consulting firm to perform a background check to verify that the candidate has not falsified any information and does not have a serious criminal record. Fitter Snacker outsources the background check because of the special skills required. If the background check is satisfactory, this information is also stored in the candidate's file, and the job offer stands. If the consultant finds evidence of falsified information or legal troubles, the file is likewise updated, and the job offer is rescinded with a written explanation.

Once the candidate passes the background check, the new employee completes additional paperwork covering employment terms and conditions, tax withholding, and benefits. All Fitter Snacker employees must sign a form that states that the employee has been given a copy of—and agrees to abide by—the company's policies and procedures. The new employee must complete an IRS W-4 form, which tells the employer the correct amount of tax to withhold from the employee's paycheck. Next, the employee must attend an orientation session, during which HR personnel describe Fitter Snacker's benefits plan. Fitter Snacker offers a comprehensive benefits plan that gives employees a range of choices on healthcare plans, life insurance, retirement plans, and medical savings accounts. The employee's dependents may also be covered under Fitter Snacker's health insurance plan. If the employee elects dependent coverage, then HR must obtain basic information about each dependent to include in the employee's file.

Because employees must provide a significant amount of detailed data to properly manage compensation and benefits, it is not surprising that Fitter Snacker frequently has problems enrolling new employees in the correct benefits plans and establishing the proper payroll deductions. It can often take a number of months before the new employee's compensation and benefits are managed correctly. The enrollment issues can generate many time-consuming phone calls to HR management—calls that would not be needed with an integrated system.

Human Resources Duties After Hiring

The Human Resources Department has responsibilities that continue beyond the hiring and job start of an employee. The HR Department should maintain a good and continual line of communication with the employee and the supervisor to make sure the employee is performing well.

Fitter Snacker, like most companies, issues performance evaluations to new and current employees. The supervisor performs an initial evaluation and reviews this evaluation with the employee. After the review, the supervisor may modify the evaluation, which both the supervisor and the employee sign. The employee may submit a written response to the review, listing any disagreements or explanations. Other senior employees, such as the plant manager, may add a separate written comment and should also review the performance evaluation and employee response. The complete package is then forwarded to the HR Department, where all documents become part of the employee's file. These files are critically important when an employee consistently fails to perform adequately. If an employee must be terminated, it is important that the company have sufficient documentation to demonstrate that the termination is warranted; otherwise the employee may file a lawsuit challenging the dismissal. Because Fitter Snacker does not have an effective information system, it is frequently difficult to manage all of the performance evaluation data. This makes it difficult for the Human Resources Department to identify problems with an employee and take corrective action (such as counseling or a transfer) before the problem leads to termination of the employee. With Fitter Snacker's paper-based system, an employee's file can be viewed by only one person at a time, and it is possible to lose track of an employee's file—temporarily or permanently. Also, it is difficult to maintain proper control of sensitive personal information when it's maintained in paper files.

Employee turnover can be a significant problem. In its 2002 Cost-per-Hire (CpH) Staffing Metrics Survey, the Society for Human Resource Management reported that hiring costs for an employee may be as high as $70,000. This figure represents both the direct costs of hiring an employee and the less tangible losses that occur during a new employee's first year or so. While employees are learning their jobs, they need training, which means that other employees have to take time from their normal jobs to train the new employee.

Another cost that is difficult to quantify is the employee's historical knowledge of the job, which is lost when an employee leaves a company. For example, if a purchasing manager leaves a company, then all of the manager's knowledge about supplier relations is lost. The company does have a record of the contract signed with the supplier, but details of the negotiations that led to the contract may not be documented. Such details can be crucial in successfully negotiating the next contract. The manager may have developed good relations with the supplier and knows who to contact when there are problems. These relationships are not specified as part of the purchasing manager position, but accrue over time with the individual holding the position. When companies experience high rates of turnover, they lose knowledge and skills that may be crucial to keeping them competitive.

Employee turnover is strongly related to job satisfaction and compensation. If employees have satisfying jobs and are well compensated, they are less likely to leave the company. Human Resources can help maintain a satisfying work environment through a number of means, including training programs for supervisors and managers, conducting periodic employee satisfaction surveys, and gathering data from employee exit surveys. Human Resources also has a critical role to play in compensation, which should be related to the skills and tasks required of the job and the performance of the employee. An important function of the HR Department is to make sure compensation levels are competitive and are applied fairly to all employees. Failure to do so can result in high rates of turnover as well as discrimination lawsuits.

ANOTHER LOOK

Boeing's Discrimination Lawsuit

Female employees first entered Boeing's workforce in large numbers during World War II to support the war effort, helping the company produce over 12,000 B-17 bombers. In spite of that early experience with women in the workforce, Boeing has had a recurring problem with pay discrimination. In 1996, the Labor Departments' Office of Federal Contract Compliance Programs (OFCCP) ran a routine investigation of Boeing's Philadelphia plant. Because of Boeing's work on federal government contracts, the OFCCP had the right to audit Boeing's level of compliance with antidiscrimination laws. It did this by comparing the relationship between the median pay of male and female employees and their median job experience. Using this median analysis, the OFCCP report stated that Boeing demonstrated "a prima facie case of systemic discrimination concerning compensation of females and minorities." Boeing's response was to conduct its own analysis. Boeing initiated the Diversity Salary Analysis (DSA) project, hiring Jon A. Geier, a partner in the Washington, D.C., law office of Paul, Hastings, Janofsky & Walker LLP, to develop a legally defensible statistical analysis of Boeing's pay practices to counter the OFCCP's median analysis. Unfortunately for Boeing, the DSA project concluded in 1997 that "gender differences in starting salaries generally continue and often increase as a result of salary planning decisions."[1] Boeing's own analysis showed that there was a pay gap for entry-level managers of $3,741.04.

In 2000, 38 women filed a class action lawsuit against Boeing, charging pay discrimination. Boeing's own salary studies supported the charges in the lawsuit, but Boeing claimed that the studies had been prepared at the direction of Boeing's lawyers and thus were protected by attorney-client privilege. While lawyers routinely prepare statistical studies to help defend a client against a lawsuit, the data contained in salary studies for business-related purposes (data that predate litigation and that are prepared by non-legal executives) do not fall under the confidentiality protection of attorney-client privilege. On October 25, 2000, Judge George Pinkie ordered Boeing to release the salary studies, explaining that "Legal departments are not citadels in which public business. . . may be placed to defeat discovery."[2]

On May 17, 2004—two days before the discrimination case was scheduled to go to trial—Boeing made a settlement offer. After negotiation, in June 2004, Boeing agreed to pay $72.5 million to settle the case.

Questions

1. Boeing settled its discrimination lawsuit for $72.5 million. Are there other costs to Boeing associated with this lawsuit? How do they compare to the financial cost of the settlement?

2. How could an integrated Human Resources information system be used to detect potential pay discrimination before it becomes systematic?

3. How would you design a compensation system so that pay discrimination is not likely to occur?

HUMAN RESOURCES WITH ERP SOFTWARE

Now that you are familiar with the numerous business processes required to manage a company's human capital, you can begin thinking about how ERP software can improve those processes, leading to overall improvements in a company's performance. With an integrated system, employee information can be stored electronically, and piles of papers and files that make the retrieval of information difficult or tedious can be eliminated. With a good information system, all relevant information for an employee can be retrieved in a matter of seconds. An integrated information system is a key component in this process.

ANOTHER LOOK

Major League Baseball Pioneers Sophisticated Data Analysis

Enterprise resource planning systems hold a wealth of data on a company—data that can be manipulated and analyzed using statistics. A wealth of data also exists on the sport of baseball, and those data are manipulated by statistics, too. Human resource managers can take a lesson from baseball when assessing their employees' performance and hiring new employees. The Society for American Baseball Research uses statistical techniques called sabermetrics. Players can be analyzed on performance attributes other than the traditional batting averages. Measures are collected on the probability of getting a hit with other players on base—for example, by determining the number of times a player has done this in the past and what type of hit the player produced. Teams such as the 2004 World Series champion Boston Red Sox are also using measures to analyze their current roster to identify weaknesses so they know which new players to recruit.

The sabermetrics method uses raw data, such as high school and college records, family backgrounds, psychological profiles, and medical histories, all culled from a wide range of sources. Teams can also measure 11 attributes including drive, endurance, leadership, self-confidence, emotional control, mental toughness, coachability, and trust.[3] Using software, baseball scouts can look for players with the potential for good fielding, or hitting, or even "hustle." Prospects are rated on physical qualities and baseball abilities such as running speed, hitting, fielding, and strength.

Companies such as Target are beginning to use this analytical approach to evaluate job applicants. Applicants complete in-store and online job applications that include questions to determine suitability for the position. These questions include true-or-false queries such as "I would rather sit around and read a book than go to a party with lots of people," and "You don't act polite when you don't want to."[4] Once the initial application is filled out, the system prompts the hiring manager with additional questions that probe more deeply into areas of concern.

Dow Chemical used to hire MBAs from Ivy League schools. The company began to realize that few of those candidates wanted to move to Midland, Michigan. These candidates also were demanding high salaries. Dow delved into the vast quantities of data in its PeopleSoft ERP system and found that its best candidates came from schools such as Michigan State, Brigham Young, and Purdue. Now Dow focuses its recruiting efforts at the

continued

institutions where it will be more successful. The cost to hire a new graduate, which exceeds $70,000, is now well spent.

Questions

1. Assume you are the HR manager at Fitter Snacker. What metrics or measurements would you develop to assess the performance of the production supervisor? The sales manager? The chief accountant?

2. How successful have the baseball leagues been in using sabermetrics? Research your answer on the Internet. Can you estimate how successful Fitter Snacker would be if they invested in a similar analysis?

Successfully using a Human Resources ERP system requires managing a significant amount of detailed information. The SAP R/3 Human Resources (HR) module provides tools for managing an organization's roles and responsibilities, definitions, personal employee information, and tasks related to time management, payroll, travel management, and employee training. Advanced HR features of SAP R/3 are discussed later in the chapter.

Most companies have an organization chart or plan that helps define management responsibilities in an organization. Without an ERP system, the organizational chart defines only the managerial relationships between employees. With an ERP system, the organizational chart provides a structure with more detail than a typical organizational chart and supports HR tasks such as recruiting employees and planning organizational changes.

SAP R/3 provides an Organization and Staffing Plan tool that is used to define a company's management structure and define the positions within the organizational structure. The Organization and Staffing Plan tool also names the person who holds each position. Figure 6-1 shows how the Fitter Snacker organizational structure could be defined in SAP R/3. The figure shows that the Fitter Snacker Company consists of three main organizational units: Manufacturing, Marketing, and Administrative. The organizational units Accounting and Human Capital Management are part of the Administrative organization. Within the Human Capital Management organization are three positions: the HCM Manager and two Analysts.

SAP R/3 distinguishes between **task**, **job**, **position**, and **person**. In SAP, an employee is a person who performs tasks, which can be assigned either to a job, which is a generic description of an employee's work responsibilities, or to the position that the person holds. Figure 6-2 shows the relationships among tasks, jobs, positions, and persons in a Marketing organization. The job of Administrative Assistant is assigned a number of tasks such as reviewing employee time charges, reviewing employee expense reports, and preparing monthly budget reports for the department. These are tasks that the company requires of any Administrative Assistant, whether that job is in Marketing, Engineering, Production, or another department. The job of Administrative Assistant can be defined once in SAP by assigning it tasks, then that definition can be used to create Administrative Assistant positions in different organizational units. The Administrative Assistant job in Marketing is one position, while the Administrative Assistant position in Accounting is a different position. Additional tasks can be added to an Administrative Assistant position to tailor it to the specific requirements of the organizational unit. For example, in Figure 6-2, the position of

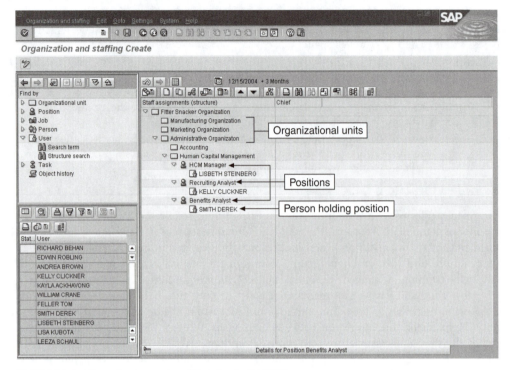

FIGURE 6-1 Organization and staffing plan in SAP R/3

Marketing Administrative Assistant has the marketing-specific task of preparing sales reports; an Administrative Assistant in Procurement would not be required to perform that task.

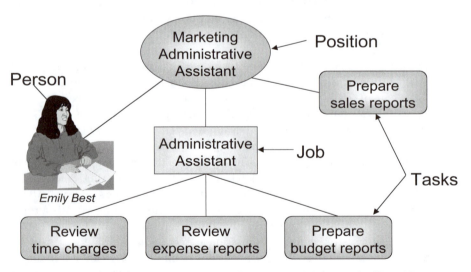

FIGURE 6-2 Relationships among positions, jobs, tasks, and persons who fill positions

Figure 6-3 shows the screen in SAP R/3 where tasks are assigned to jobs—in this case, the task Prepare Budget Report is assigned to the job Administrative Assistant. If the tasks associated with jobs and positions are well defined and current, then the recruiting process is simplified because it is easier to determine whether candidates have the qualifications required for the job. In addition, determining appropriate compensation for a position is simplified if the tasks required for each position in a company are clearly and consistently defined.

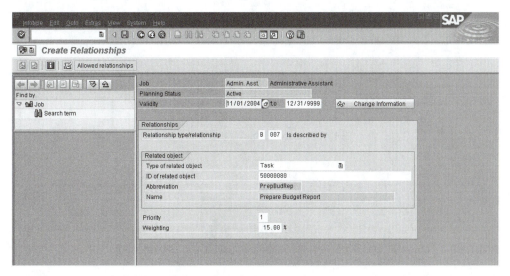

FIGURE 6-3   Assignment of a task to a job in SAP R/3

Complete and accurate human resource data simplify a manager's duties. The Manager's Desktop tool within the SAP HR module provides access to all the Human Resource data and transactions in one location. Figure 6-4 shows the Personal Data portion of the Manager's Desktop. This area provides all of the data maintained in the Human Resources module for all employees who report to the manager. Human Resource data is very sensitive because it is related to employees' personal information, so controlling access to it is critical. An advantage of an integrated information system over a paper-based system is that controlling access to data is automated; managers can use the system to determine which users should have access to various data.

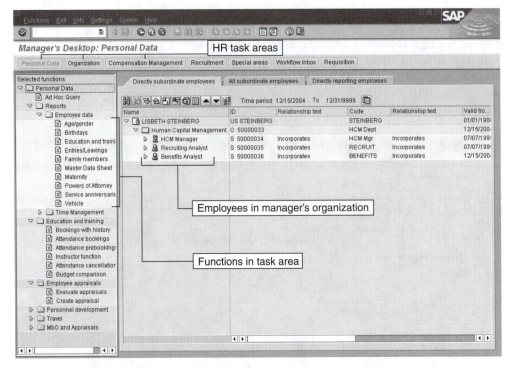

FIGURE 6-4 Manager's Desktop provides single-point access to HR functions

ADVANCED SAP R/3 HR FEATURES

Discussing in detail the many processes related to Human Resources is beyond the scope of this text; however, some of the advanced features of the Human Resource Module in the SAP R/3 system are discussed below.

Time Management

Hourly employees, who are paid for each hour worked, must record the time that they work so that they can be paid. While salaried employees are not paid based on hours worked, their time must usually be tracked as well. Any time not worked must be attributed to vacation or leave. In addition, for cost accounting purposes, it is important to be able to attribute an employee's time to a cost object (such as cost center, project, or production order). The SAP R/3 system uses Cross Application Time Sheets (CATS) to record employee working times and provide the data to applications that include:

- The SAP Controlling Module, for cost management
- The SAP Payroll module, to calculate employee pay, which is transferred to the Financial Accounting Module
- The SAP Production Planning Module, to determine whether enough labor is available to support production plans

Payroll

Payroll is probably the most important Human Resources function. Employees are, not surprisingly, very particular about being paid the correct amount at the correct time! Many people live paycheck-to-paycheck, meaning that getting paid correctly and on time is crucial. Without proper management of the payroll process, employees might not be paid for all of the hours they worked, they might not be paid at the appropriate rate, or they might have too much or too little money withheld from their pay for taxes and benefits. Mistakes in payroll can cause significant job dissatisfaction. The two key processes in determining the pay an employee receives are calculation of the remuneration elements and determination of statutory and voluntary deductions. The **remuneration elements** of an employee's pay include the base pay, bonuses, gratuities, overtime, sick pay, and vacation allowances that the employee has earned during the pay period. The **statutory and voluntary deductions** include taxes (federal, state, local, Social Security, and Medicare), company loans, and benefit contributions. Properly determining the pay for an employee requires accurate input data and evaluation of remuneration elements and deductions.

The process of determining each employee's pay is called a **payroll run**. In the payroll run, the SAP R/3 system evaluates the input data and notes any discrepancies in an **error log**. Payroll employees review the error log, make the necessary corrections, and repeat the payroll run until no errors are recorded. Then the payroll run is used to generate information for accounting, electronic funds transfers, employee pay statements, withholding tax payments, and other calculations.

Travel Management

Companies can spend a significant amount of money on employee travel, and managing travel and its associated expenses can be a significant task. A travel request, which may originate with the employee or the employee's manager, is the first step in the travel management process. Travel requests usually require management approval, and the level at which travel must be approved may depend on the duration and location of the travel. Once the travel is approved, travel reservations must be made. Because airfare, hotel, and rental car costs can vary widely, companies frequently require employees to make reservations through either a company travel office or travel agency under contract to the company. The employee must keep receipts for travel expenses incurred during the trip in order to complete an expense report and receive reimbursement. The SAP R/3 Travel Management system facilitates this process by maintaining travel data for each employee, including flight, hotel, and car preferences and integrating this data with the Payroll module (for reimbursements) and with the Financial Accounting and Controlling modules to properly record travel expenses. Submitting an expense report can be simplified using a Web-based application that allows employees to submit reports through a Web browser.

153

ANOTHER LOOK

FXIS: From SAP Customer to Solution Provider

Photocopier manufacturer Fuji Xerox established Fuji Xerox Information Services (FXIS) in 1984 to create software for its computer-related products. Since that time, FXIS has taken over responsibility for managing Fuji Xerox's information systems and networks, and has expanded its business to include sales of computer and network equipment as well as education and training in computer software applications.

In 1998, FXIS sought to improve its own internal software systems with the goal of being able to close its books (reconcile all of its accounting records) two days after the end of the month. To accomplish this task, FXIS chose to implement SAP R/3.

When planning its SAP implementation, FXIS decided that to meet its goal of closing its books in two working days it should use SAP for all of its business applications and get rid of legacy systems. Unlike some companies using SAP, FXIS emphasizes the integration of HR data. FXIS implemented the Sales and Distribution, Materials Management, Financial Accounting, Controlling, Project System, and Human Resources modules of the R/3 system. To meet the two-day closing requirement, FXIS needed to have expense account information from its sales force and time charge information from its software developers and system administrators. To simplify the collection of travel and time charge data, FXIS created its Direct Input (DI) system, which makes it easy for users to enter their time charge data into the system through a Web interface. The DI system stores this data in its own database, which allows managers to review and approve the data. When the data have been approved, the information is transferred to the SAP R/3 system.

The DI system provides the advantage of a familiar browser interface for users who may not be comfortable using SAP R/3 directly. It also does not require additional software or hardware. The DI system has proved so successful that FXIS has formed an ERP solution business and has developed standard templates for its DI system, which it is marketing to other SAP customers in Asia.

Questions

1. What are the advantages and disadvantages of the FXIS decision to use SAP for all business processes and eliminate its legacy system?

2. How does the use of the DI system contribute to FXIS's goal of closing its books in two days? Do you think there are any problems created by not having the data collected by the DI system available in real time?

Training and Development

The Personnel Development component of the SAP R/3 Human Resources module supports the planning and implementation of employee development and training activities. Such education maximizes an employee's ability to contribute to the organization. Employee development is driven by qualifications and requirements. **Requirements** are skills or abilities associated with a position, while **qualifications** are skills or abilities associated with a specific employee. Requirements and qualifications refer to the same concept from a different perspective. Using the Personnel Development tool, an employee's qualifications can be compared with the requirements for a position to which the employee aspires. Comparing an employee's qualifications with the position's requirements allows a manager to identify gaps and plan development and training efforts to close the gaps. This comparison can also serve as a basis for employee evaluation, and can motivate the employee by providing a goal and the means to achieve it.

One of the most important reasons for managing the development and training of employees is **succession planning**. The success of a company depends in large part on the skills, abilities, and experience of its management team. This is especially true for a small company like Fitter Snacker. The company should have a plan for replacing key employees when they leave the company. Savvy customers have been known to avoid establishing long-term relationships with companies that do not have well-developed succession plans.

ANOTHER LOOK

Management Succession Planning at Quest Diagnostics

Finding a chief executive officer (CEO) can be a tremendous challenge for a company. For many companies—including Coca-Cola, Xerox, and Procter & Gamble—finding a new CEO has been a process marked by long searches, poor choices, and fumbled transitions. In other cases, succession planning has worked to ensure a smooth transition. When McDonald's Corporation CEO James Cantalupo died of a heart attack on April 19, 2004, succession planning allowed the board of directors to name Charles H. Bell as the new CEO within hours. Unfortunately, McDonald's succession planning capabilities were quickly tested a second time when Bell was diagnosed with cancer only one month after being named CEO. Bell was replaced by Jim Skinner in November, 2004. Bell died January 17, 2005.

At Quest Diagnostics, CEO Kenneth W. Freeman worked for five years to prepare Surya N. Mohapatra to take over the CEO position on May 4, 2004, the date Freeman had set for his own retirement. Experts say Freeman's approach is an unusual case study in how to choose and prepare a future CEO.

In 1999, Freeman was managing a series of acquisitions that would triple Quest's revenues in five years. But he knew that the pace of acquisitions would slow eventually and that Quest's next CEO would need a science background to generate growth internally. First, Freeman identified candidates by putting 200 Quest executives through daylong case assignments to evaluate their leadership skills. From this screening process, one executive stood out—his new chief operating officer, Surya N. Mohapatra. Mohapatra had extensive experience in cardiovascular disease and information technology—important areas for Quest. Freeman gave Mohapatra a clear (but by no means guaranteed) shot at the top job and began the task of preparing him to become CEO. While Mohapatra had many positive skills and abilities, he also had areas that needed improvement. One of these areas was public speaking. Freeman decided the best method to develop Mohapatra's speaking skills was trial by fire. As part of his development, Freeman had Mohapatra make unscripted comments to employees, meet with shareholders, and field questions from analysts on conference calls. As a result, Mohapatra has become a more polished and confident speaker.

continued

Mohapatra's background as a scientist probably contributed to habits that Freeman felt could undermine him as a CEO. Mohapatra sometimes took weeks to make decisions that should only take days, and was far more "hands-on" than Freeman believed he should be. For example, Mohapatra would sometimes reopen interviews for jobs that his subordinates were ready to fill. Freeman challenged Mohapatra to make faster decisions and give his executive team more authority. For five years, the two engaged in lengthy telephone conversations during which Freeman would analyze Mohapatra's evolving management style and suggest further improvements. Freeman now concedes that the training was "pure browbeating," but it worked. "Am I more ready now than I was four years ago? Absolutely," says Mohapatra.

An important part of the CEO position is managing the relationship with the company's board of directors. In the beginning, Mohapatra would naturally defer to Freeman in board discussions. Freeman worked to make Mohapatra more assertive. At first he did this surreptitiously, by leaving the room during discussions, and later had him conduct formal board presentations. To increase Mohapatra's comfort level with the directors, Freeman changed the seating chart to increase Mohapatra's face time with other directors.

Freeman believed that when a business outgrows its leader, as Quest did, the CEO should relinquish power. "Being a CEO is not a birthright," he says. "It's a privilege." Mohapatra's success is far from assured. But Freeman did as much as possible to make sure that Quest would have a CEO prepared to handle the challenges of corporate leadership.[5]

Questions

1. If you were developing a succession plan for a key position at a company, how would you screen employees to determine potential candidate(s)? Would you select more than one employee for development?

2. Can identifying a successor for the CEO position in advance cause problems in an organization? If so, what might these problems be, and how might you handle them?

NEW HUMAN RESOURCES FEATURES OF SAP R/3

To keep pace with rapid changes in the social and legislative developments in the corporate world, the SAP R/3 HR module has been expanded to include features to assist in managing some HR tasks that have only recently become important to corporations.

Mobile Time Management

Many employees, especially sales personnel who spend a significant amount of time on the road, may not have regular access to a PC. Mobile Time Management allows employees to use cellular phones to record their working times, record absences, enter a leave request, and check their time charge data.

Management of Family and Medical Leave

The Human Resources module reduces the administrative burden imposed by the federal Family and Medical Leave Act (FMLA) of 1993. The HR system can now determine whether an employee is eligible to take FMLA absences and automatically deducts those absences from the days the employee takes from allowable leave.

Domestic Partner Handling

Many companies provide benefits for domestic (unmarried) partners. The Human Resources module now supports the management of benefits for domestic partners and their children. The system now provides more flexibility in customizing dependent coverage options for health plans, eligibility for enrollment of dependents, and designation of beneficiaries.

Administration of Long-Term Incentives

An outgrowth of the Sarbanes-Oxley Act (see Chapter 5) is that companies must account for the expected costs that occur as a result of long-term incentives such as the exercising of stock options. The Human Resources module now provides more options for processing long-term incentives. Integration with the SAP Payroll module enables companies to calculate taxes accurately when employees exercise incentives and sell their shares in the company. SAP can share the incentive data with Accounting so that Accounting can do the necessary reporting.

Personnel Cost Planning

Changes in an organization (including expansions, acquisitions, and downsizing) can have an impact on employee-related expenses, which are usually a significant portion of a company's costs. The Personnel Cost Planning tool allows HR personnel to define and evaluate planning scenarios to generate cost estimates. Cost planning and simulation allows HR to forecast cost estimates by integrating data with other SAP R/3 modules.

Management and Payroll for Global Employees

The management of global employees involves many complicated issues, including relocation plans, visa and work permits, housing, taxes, and bonus pay. SAP R/3 has enhanced features to support the management of these issues with customized functionality for over 50 countries, so that payroll processes meet current legal regulations and collective bargaining agreements in the local business environments.

Management by Objectives

The concept of management by objectives (MBO) was first outlined by Peter Drucker in his 1954 book *The Practice of Management*. In MBO, managers are encouraged to focus on results, not activities, and "negotiating a contract of goals" with their subordinates without dictating the exact methods for achieving them. SAP R/3 provides a comprehensive process to support the MBO approach that incorporates performance appraisal. The appraisal results can affect the employee's compensation, generating annual pay raises that can be either significant or insignificant, depending on the employee's performance. Likewise, managers can include the results of achieved objectives in the employee's qualifications profile.

HR Management at the U.S. Postal Service

The United States Postal Service signed a deal with SAP in August 2004 to provide human resource management software to replace its existing information system software, which was written in the 1980s. The USPS is one of the largest public-sector organizations in the world, employing roughly 70,000 workers, which is about one-third of the civilian government workforce. The USPS chose SAP in part because SAP has experience in providing HR systems to about 50 other postal services worldwide. The deal is worth an estimated $35 million, with $14 million for software and $21 million for consultants.

Postal service managers expect the SAP software to help them better control costs and manage employees. In addition, the managers will be able to view information company-wide and allocate resources as needed. Robert Otto, chief technology officer for the USPS, says, "This project will enable us to modernize and centralize our human capital management capabilities and help us provide improved access to information for our employees through an integrated IT platform. The desired result of the project is to leverage technology to provide more value to our most valuable asset, our employees."[6]

Questions

1. What human resource management problems are unique to the USPS? How will an ERP system like SAP help solve those problems?
2. What recommendations do you have for the USPS in its implementation of this large system? Present you ideas in a memo to the CIO of the USPS.

Chapter Summary

1. Employees are among a company's most important assets. Without qualified and motivated employees, a company cannot succeed.

2. The Human Resources Department has the primary responsibility of ensuring that the company can find, evaluate, hire, develop, evaluate, and compensate the right employees to achieve the company's goals. HR is also responsible for employee training and development, succession planning, and termination.

3. Managing, sharing, controlling, and evaluating the data required to manage a company's human capital is simplified with an integrated information system.

4. Effective Human Resource processes require a system that controls the flow of timely and accurate information within and among departments.

Key Terms

Error log

Human capital management (HCM)

Job

Payroll run

Person

Position

Qualifications

Remuneration elements

Requirements

Short list

Statutory and voluntary deductions

Succession planning

Task

Exercises

1. Describe a position in a company that you would like to have. List the tasks that you think would be required for this position.

2. Suppose you are designing a system to summarize information from resumes submitted to a company's Human Resources Department. Create a list of the information that you think would be useful to collect from the resumes.

3. Describe a position in a company that you would like to have after five years of work experience. List the requirements that you think would be required to hold this position. List the qualifications that you currently possess. Describe how you plan to obtain the qualifications necessary to hold the position.

4. Suppose you are a manager of Fitter Snacker's Sales Department. What Human Resources information do you think you would need to manage your sales force?

5. List the steps in a typical recruiting process. Highlight the steps that involve interaction with the potential job candidate. Identify problems in the process that might lead a candidate to develop a negative opinion of the company. How might an effective information system reduce the potential for these problems? Incorporate into your answer experiences you may have had in looking for a job.

End Notes

[1] Stanley Holmes and Mike France, "Coverup At Boeing? Internal documents suggest a campaign to suppress evidence in a pay-bias lawsuit," *Business Week* (June 28, 2004).

[2] Holmes and France, "Coverup At Boeing? Internal documents suggest a campaign to suppress evidence in a pay-bias lawsuit."

[3] CNET.com, "Postal Service seals big SAP deal," http://news.com.com (August 23, 2004); Internetnews.com "SAP Delivers for the Mailman," http://www.internetnews.com/ent-news/article.php/3399361 (August 24, 2004); http://www.sap.com, "US Postal Service Selects mySAP Business Suite," *SAP & Partner News* (August 28, 2004).

[4] Mel Duvall, "Boston Red Sox: Backstop Your Business," *Baseline*, http://www.baselinemag.com (May 14, 2004).

[5] Duvall, "Boston Red Sox: Backstop Your Business."

[6] Louis Lavelle, "How to Groom the Next Boss," *Business Week* (May 10, 2004).

160

CHAPTER **7**

PROCESS MODELING, PROCESS IMPROVEMENT, AND ERP IMPLEMENTATION

LEARNING OBJECTIVES

After completing this chapter, you should be able to:

- Use basic flowcharting techniques to map a business process.
- Develop an Event Processing Chain (EPC) diagram of a basic business process.
- Evaluate the value added by each step in a business process.
- Develop process improvement suggestions.
- Discuss the key issues in managing an ERP implementation project.
- Describe some of the key tools used in managing an ERP implementation project.

INTRODUCTION

The underlying theme of this text is the management of business processes. In this chapter, we explore

tools that can be used to describe processes, which can then be improved. We finish by describing the role

these process modeling tools play in ERP implementation projects.

PROCESS MODELING

By now, it should be clear that business processes can be quite complex. Individuals with various skills and abilities are responsible for executing business processes. If the business processes are not clearly defined and if the individuals are not adequately trained in their roles and in how their roles fit within the business process, it is unlikely that the business process will be effective (achieve the desired results) or efficient (achieve the desired results with the minimum use of resources). A range of terms, including flowchart and process map, have been used to describe methods of representing processes. We will use the term **process model** to describe any abstract representation of a process. A process model can be as simple as a diagram with boxes and arrows or as complex as a graphical representation in computer software that allows for simulation of the process. Process modeling tools provide a way to describe a business process so that all participants can understand the process. Frequently, process models are developed by a team of employees involved in the process, and the process of developing the model frequently reveals misunderstandings and ensures that all team members are "on the same page." Graphical descriptions are usually easier to understand, and a well-developed process model provides a good starting point for analyzing the process so that improvements can be designed and implemented. Process models also document the business process so that employees can more easily be trained on how to support the business process.

Flowcharting Process Models

Flowcharts are the simplest of the process modeling tools. A **flowchart** can be defined as any graphical representation of the movement or flow of concrete or abstract items—materials, documents, logic, and so on. Flowcharts originated with computer programmers and mathematicians, who used them to trace the logical path of an algorithm. In the early days of computer programming, computer resources were limited and executing a program used considerable resources. As a result, most programmers spent a considerable amount of their time clearly defining the logic of their programs using flowcharts before actually writing the code and testing the program.

A flowchart is a clear, graphical representation of a process from beginning to end, regardless of whether that process is an algorithm or a manufacturing procedure.[1] Flowcharting has been used since the 1960s in business applications to help businesspeople visualize workflows and functional responsibility within organizations. Today the term **process mapping** is used interchangeably with flowcharting, the distinction being that process mapping specifically targets the activities occurring within an *existing* business process. Process mapping develops an "as-is" representation of the process, with a goal of exposing weaknesses that need to be addressed. Once a process map has been developed, a **gap analysis**, which is an assessment of disparities between an organization's current situation and its long-term goals, can be performed.

Flowcharting uses a standardized set of symbols to represent various business activities. Few symbols are required to define a business process. Figure 7-1 shows the basic flowcharting symbols. You can use a wide range of symbols for process mapping, but the basic set shown in Figure 7-1 is sufficient to describe even a complicated business process. Using a few, simple symbols places the focus on the process and not the tool used to represent it.

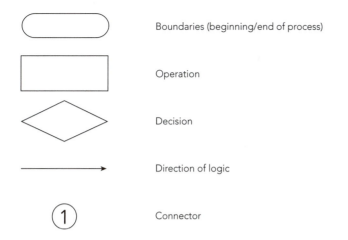

	Boundaries (beginning/end of process)
	Operation
	Decision
	Direction of logic
①	Connector

FIGURE 7-1 Basic flowcharting symbols

The following description of the process Fitter Snacker uses to reimburse sales personnel for their business expenses illustrates process mapping.

Fitter Snacker Expense Report Process

After a Fitter Snacker salesperson incurs travel expenses on her credit card, she fills out a paper expense report, makes a copy for her records, attaches receipts for any expenses over $25, and mails it to her zone manager at the branch office. The manager reviews it and either approves the report or mails it back to the salesperson with a note asking for an explanation, verification, or modification. Once the manager approves the expense report, it is mailed to the corporate office. After the mail is sorted at the corporate office, the expense report is forwarded to the accounts payable (A/P) clerk, who performs a preliminary check of the report. The clerk contacts the zone manager for any necessary clarification. The clerk then forwards the expense report to the expense report auditor, who reviews the expense report. If there is a problem with the report, the auditor mails it back to the salesperson, who revises and returns it. Then the auditor enters the report into Fitter Snacker's PC-based accounting system and files a hard copy with the receipts in a filing cabinet, organized by employee name.

At the end of each week, an A/P clerk uses the PC-based accounting system to print payroll checks, payments to suppliers, and expense reimbursement checks. When the salesperson receives her reimbursement check, she deposits it into her checking account and mails a payment to the credit card company, which credits the employee's card account. Figure 7-2 shows the process map for the first part of the current Fitter Snacker expense reporting process.

One of the most important decisions to make in process mapping is to define the process boundaries. The **process boundaries** define which activities are to be included in the

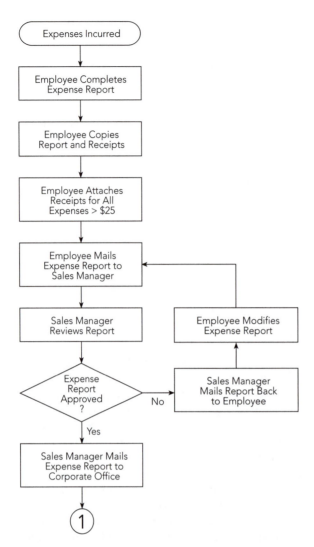

FIGURE 7-2 Partial process map for Fitter Snacker expense reporting process

process, and what is considered part of the environment—external to the process. It is important to clearly define which activities are part of the process and which are external to the process. Many activities related to sales employee expenses are considered outside of the boundary of the process map shown in Figure 7-2. For example, a company might provide cash advances or issue corporate credit cards before travel expenses are incurred. Employees might need to make reservations for air travel or hotels through a corporate travel office. The company might want employees to use a preferred hotel, or might have specific policies regarding car rentals, including preferred rental companies, approved car classes, insurance, and prepaid gas. While all of these considerations are important and should be documented somewhere, the process mapped in Figure 7-2 is the *expense reporting process*; and the process boundaries do not include these additional activities.

The process map begins after travel expenses have been incurred and ends with the receipt by the salesperson of a refund check. These starting and end points are the process boundaries for the expense reporting process.

In Figure 7-2, the beginning process boundary is represented by the oval figure containing the text "Expenses Incurred." All processes should have only one beginning point and one ending point. After the Expenses Incurred oval, the process map shows four operation blocks that define the tasks performed by the employee in completing the expense report. The choice of including four operation blocks and the level of detail in the descriptions is a matter of the user's preference and depends on the purpose for which the process map is created. If the process map is to be used to improve a business process and the members of the process improvement team are familiar with the process, then less detail is needed. In fact, too much detail could obscure the key features of the process. On the other hand, the process map might be used to document the process for training new sales employees. In that case, more detail is needed so that new employees use the process map to follow the process properly.

Figure 7-2 also contains a decision diamond. A decision diamond asks a question that can be answered with "yes" or "no." In the figure, the decision diamond asks whether the Sales Manager approves the expense report. There are only two possible options—yes or no. It is tempting for the novice to create process maps with decision diamonds that have more than two outcomes. Doing so can lead to confused logic—all business processes can be defined using one or more decision diamonds, each with only two outcomes.

Finally, because Figure 7-2 only shows part of the expense reporting process, the flowchart ends with a connector. Most business processes are too complicated to fit on a single sheet of paper. Connectors provide a way to continue process maps from one sheet to the next.

Exercise 7.1

Complete the process map for the Fitter Snacker expense report process started in Figure 7-2 using the process mapping symbols shown in Figure 7-1.

Extensions of Process Mapping

The development of computer technology, specifically high-quality graphical interfaces, has allowed process mapping tools to evolve beyond the simple symbols of flowcharting. One helpful tool is **hierarchical modeling**, which is the ability to flexibly describe a business process in greater or less detail depending on the task at hand. A hierarchical model of Fitter Snacker's expense report process is illustrated in Figure 7-3.

It is important to document (for training purposes) the detailed steps that the salesperson follows to complete the expense report, but the details can make the process map cumbersome for process improvement activities. In hierarchical modeling, the four steps that the salesperson follows to complete the expense report can be condensed into one step, designated as Employee Prepares Expense Report. Modeling software that supports hierarchical modeling provides the user with the flexibility to move easily from higher-level, less detailed views to the lower-level, more detailed views. Hierarchical modeling can aid in process mapping by allowing a user to create a broad, high-level view of a process and then add more detail as the process is analyzed.

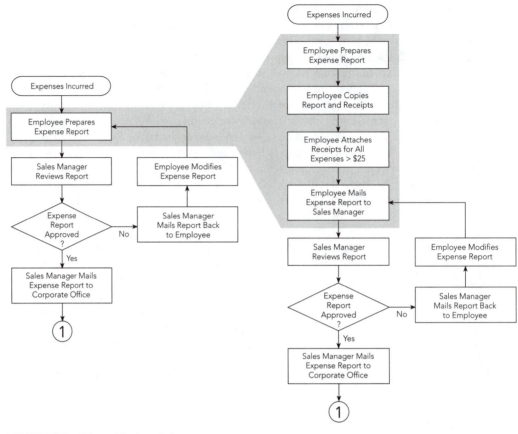

FIGURE 7-3 Hierarchical modeling

Another widely used and widely recognized type of process mapping technique is **deployment flowcharting**. This type of flowchart depicts team members across the top, with each step aligned vertically under the appropriate employee or team. Figure 7-4 shows the Fitter Snacker expense report process as a deployment flowchart. This type of process map is also referred to as a **swimlane flowchart**. This process mapping technique has the advantage of clearly identifying each person's tasks in the process.

Event Process Chain (EPC) Diagrams

ERP software such as SAP consists of business applications that support business processes. SAP's software supports hundreds of business processes, and SAP has developed graphical models for many of these business processes using the **event process chain (EPC)** format. The EPC format uses only two symbols to represent a business process. The advantage of the EPC format is that it matches the logic and structure of SAP's ERP software design. The EPC modeling technique is available as a software tool through the IDS/Scheer company as the Aris (Architecture of Integrated Information System) toolset.

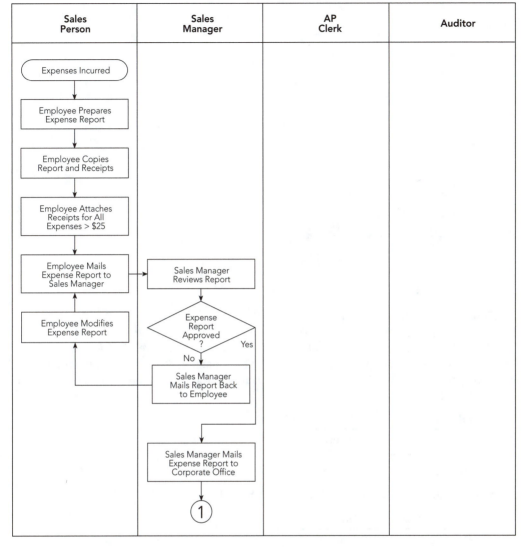

Sales Person	Sales Manager	AP Clerk	Auditor

FIGURE 7-4 Deployment, or swimlane, flowcharting

The two structures used in EPC modeling to represent business processes are events and functions. As shown in Figure 7-5, events reflect a state or status in the process and functions represent the part of the process where change occurs.

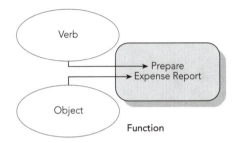

FIGURE 7-5   EPC components

Unlike flowcharting, EPC modeling enforces a strict structure. EPC software enforces an event-function-event structure. A standardized naming convention for functions and events is also used in EPC modeling. For example, the convention for naming events is Object→Past Participle:

Object	Past Participle
Expense	Incurred
Expense report	Approved
Hard copy	Filed

For functions, the naming convention is Verb→Object:

Verb	Object
Prepare	Expense report
Review	Expense report
Mail	Refund check

Figure 7-6 shows a simple EPC diagram for part of the Fitter Snacker expense report process. EPC diagrams begin and end with events. Furthermore, events must be followed by functions, and functions must be followed by events.

In addition to direct connection of events to functions, EPC diagrams employ three types of branching connectors. Branching occurs when logic either comes from more than one source or proceeds to more than one potential outcome. The three connector types are AND, OR, and Exclusive OR (XOR). Figure 7-7 shows an application of the OR connector, which indicates that after the payment is processed, the salesperson is notified, or the sales manager is notified, or both the salesperson and sales manager might be notified.

FIGURE 7-6   Basic EPC layout

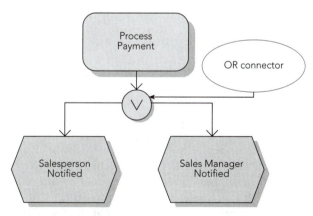

FIGURE 7-7   OR connector

169

Process Modeling, Process Improvement, and ERP Implementation

Figure 7-8 shows how the AND connector functions. In the figure, the expense report must be recorded and a hard copy must be filed.

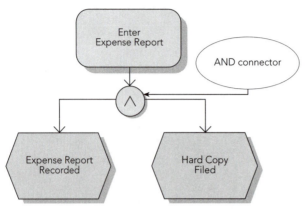

FIGURE 7-8   AND connector

Figure 7-9 shows how the XOR connector can be used to represent the manager's decision on whether to approve the expense report. The XOR connector is exclusive, so in Figure 7-9, only one event can occur after reviewing the expense report: it is approved or it is not approved.

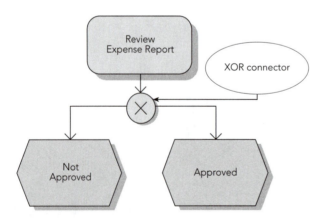

FIGURE 7-9   XOR connector

Figures 7-7 through 7-9 show one function connecting to two events using a branch connector, but it is not always the case that functions lead into the connector, nor do multiple events always follow the connector. For example, Fitter Snacker could require that salespeople complete expense reports at the end of a sales trip, but trips lasting more than one week require that the salesperson complete an expense report at the end of each week. This condition is illustrated in Figure 7-10, where the preparation of the expense report can be triggered by either of two events: the end of the trip *or* the end of the week. It is also possible that more than two events could trigger the function.

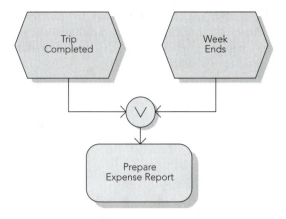

FIGURE 7-10   OR connector with two triggering events

Figure 7-11 shows all possible connection combinations. Note that it is not possible to have a single event connect to multiple functions with OR or XOR connectors, because events represent a status or state. Because OR and XOR connectors require a decision, they must be preceded by a function so that a decision can be made.

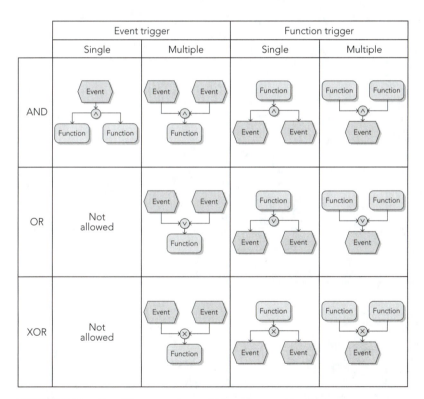

FIGURE 7-11   Possible connector and triggering combinations

Process Modeling, Process Improvement, and ERP Implementation

Finally, Figure 7-12 shows splitting and consolidating of a path through the process. In this case, the Fitter Snacker salesperson can submit her expense report online if she has Internet access; otherwise, she must send in a paper report. Note that the type of branch connector that splits the path also must be used to consolidate it.

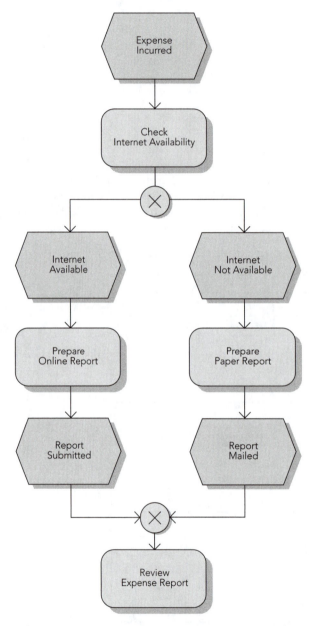

FIGURE 7-12   Splitting and consolidating process paths

The basic EPC diagram can be complemented with additional information. For example, Figure 7-13 shows the first part of an EPC diagram for the Fitter Snacker expense report process that also shows data elements (unapproved multi-copy expense reports) and organizational elements (salesperson, sales manager). The additional elements allow for a more complete description of the process, documenting the "who" and "what" aspects of the process.

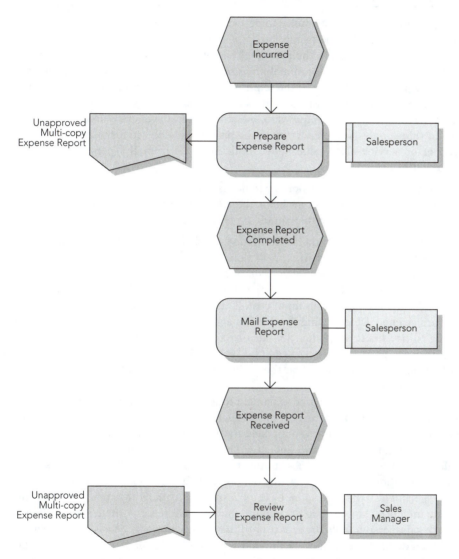

FIGURE 7-13   EPC diagram with organizational and data elements

Exercise 7.2

Using the above description, complete the EPC diagram of the Fitter Snacker Expense Report process shown in Figure 7-13. Add both the logic elements (events, functions, and connectors) as well as the organizational elements.

PROCESS IMPROVEMENT

Process mapping tools provide the ability to describe business processes in a universally understood format. Generally, the task of mapping a process requires a team consisting of key personnel who are involved in the process. The act of accurately describing the process and understanding how the functional areas interact frequently makes the necessary steps to improve processes obvious to the team. This is especially true for organizations that have focused on functional responsibilities and not on business processes.

The simple technique of value analysis can also be used to generate process improvement ideas. In **value analysis**, each activity in the process is analyzed for the value it adds to the product or service. The value added is determined from the perspective of the customer. Activities can add:

- *Real value*: Value for which customer is willing to pay
- *Business value*: Value that helps the company run its business
- *No value*: An activity that should be eliminated

Activities that cost more than their value added should be improved. The Fitter Snacker expense report process does not provide real value, because Fitter Snacker's customers do not care whether sales employees receive prompt and accurate reimbursement of their business expenses. However, the expense report process does provide business value, and it should provide this value at a minimum cost. Evaluating the value of a business activity is not a hard science. Determining the value of a good or service is easy—it's what someone is willing to pay for it. Applying this idea to a part of a business process is more challenging, because parts of a process can't be "sold" on the open market. While a challenging task, evaluating each activity on the basis of value provided can highlight opportunities for improvement.

The value analysis concept can be expanded to look at both the time and cost of each process step. For each step in the current process, you would estimate the actual time and cost. Then you would estimate the value-added time and cost—determining how much of the actual time is adding value and how much of the cost is worth paying for.

We will use a Fitter Snacker process to illustrate value analysis. The company's *mail expense report* function could cost upwards of $50, including not just the cost of the envelope and postage, but also the time spent by the salesperson to mail the expense report. The value analysis includes elapsed time for mailing the expense report—the length of time from when the salesperson mails the report until the sales manager receives it. This elapsed time should include the time it takes for the salesperson to find a mailbox, the time for the postal service to deliver the expense report to the company headquarters, plus the time it takes the company's internal mail system to deliver the expense report to the sales manager.

Suppose that for Fitter Snacker the elapsed time is typically three days. To perform value analysis, you would determine how much of this time and cost is value-added. To determine this value, you must view the activity *mail expense report* in terms of what is actually being accomplished. It is not the physical transmission of paper that matters. Mailing the expense report is a means to transmit expense data and documentation. E-mail minimizes the time and cost required to transmit data; therefore, mailing an expense report at a cost of $50 to arrive in three days is only providing value worth pennies (which is the cost to send an e-mail) in a function that should be able to be executed within seconds. Looking at the current time and cost of each process step, then asking "what is actually being accomplished and what is the value?" can help identify areas for process improvement.

Each step in a business process should be challenged to determine if it is providing value. H. James Harrington, in his book *Business Process Improvement*, suggests that companies ask the following questions about their business processes to identify areas for improvement:

- Are there unnecessary checks and balances?
- Does the activity inspect or approve someone else's work?
- Does it require more than one signature?
- Are multiple copies required?
- Are copies stored for no apparent reason?
- Are copies sent to people who do not need the information?
- Is there unnecessary written correspondence?
- Are there people or agencies involved that impede the effectiveness and efficiency of the process?
- Do existing organizational procedures regularly impede the efficient, effective, and timely performance of duties?
- Is someone approving something they already approved (for example, approving capital expenditures that were approved as part of a budget)?
- Is the same information being collected at more than one time or location?
- Are duplicate databases being maintained?

Harrington also suggests concepts that can be used to improve processes:

- Perform activities in parallel, for example, approvals.
- Change the sequence of activities.
- Reduce interruptions.
- Avoid duplication or fragmentation of tasks.
- Avoid complex flows and bottlenecks.
- Combine similar activities.
- Reduce the amount of handling.
- Eliminate unused data.
- Eliminate copies.[2]

Evaluating Process Improvement

While identifying process improvements is challenging, implementing them is even more challenging. Disrupting the current process and to make changes can be costly and time consuming, and managers are frequently reluctant to risk trying process improvement ideas, especially if the ideas require significantly different ways of doing things. One way to overcome this risk is to use dynamic process modeling to evaluate process changes before they are implemented. **Dynamic process modeling** takes a basic process flowchart and puts it into motion using computer simulation techniques to facilitate the evaluation of proposed process changes. Computer simulation uses repeated generation of random variables (such as customer orders) that interact with a logical model of the process to predict the performance of the actual system. These models can estimate the performance of the system using measures such as cycle time (how long the process takes), productivity, total cost, idle time, and bottlenecks.

ANOTHER LOOK

Business Process Innovation at Nova Chemicals

Nova Chemicals has used business process innovation to move from a function-oriented company to a process-oriented company (see Chapter 1). According to John Wheeler, CIO of Nova Chemicals, "**Business process innovation (BPI)** is the process of improving processes. BPI is based on understanding the way you work. Once you understand the way you work, you can begin to improve the way you work." As part of its BPI initiative, Nova has used IDS/Scheer's Aris toolset to document its business processes. Nova has seen the Aris toolset as a "huge enabler" in BPI. The tool requires structure and discipline to use, but allows companies to understand all of its processes, not just the workflow. Wheeler estimates that in the early days of computer information systems, technology was 75 to 80 percent of the cost of a project, while today technology is only 10 to 15 percent of the project's cost. Wheeler estimates 30 to 40 percent of the budget for a business process improvement project is spent on understanding the current process.

Nova has seen considerable success using BPI in its sales process. In this project, Nova gathered top salespeople and had them document the sales process using the Aris toolset. They discovered that every salesperson performed the process differently, and in facilitating innovation sessions, they were able to develop the best practices for use throughout the company, taking the best ideas from each participant. Wheeler has observed that thought leaders—the people who contribute most to generating process improvement ideas—can come from anywhere in the organization and it is frequently surprising who the thought leaders turn out to be.

Wheeler sees BPI as just the next step in the evolution of process improvement methods. Other methods for improving processes include quality circles, continuous improvement, and business process reengineering. In many ways, BPI resembles quality circles, the quality improvement technique pioneered in Japan in the 1970's. In **quality circles**, employees in a department have regular team meetings to discuss problems and collaboratively develop solutions. Quality circles were followed by the **continuous improvement** quality philosophy, again imported from Japan, which prescribed systematic and repeated improvement efforts. The concept of **business process reengineering** was popularized by Michael Hammer and James Champy in their 1993 book *Reengineering the Corporation*. Unlike continuous improvement, which stressed

continued

repeated improvement activities to achieve gradual improvements, business process reengineering recommends radical change to achieve radical improvements. BPI is the latest evolution of improvement methodologies.

Some people might see the continuously changing methods merely as fads that allow consultants to sell new consulting services. Wheeler doesn't see it that way. Innovations in methods and techniques in process improvement can keep innovation fresh and new. To Wheeler, BPI is itself a process, and all processes can be improved.

Questions:

1. What role does process mapping play in business process improvement? Is there an advantage to using a structured approach like IDS/Scheer's Aris toolset?
2. How would you form a team to improve a business process? How would you manage the project? What tools would you use?
3. Can you think of any ways in which technology can contribute to further improvements in the process of improving business processes?

ERP WORKFLOW TOOLS

Most business processes are performed regularly, enabling employees responsible for the process to become efficient in the tasks involved in the process. For example, the sales order process is fundamental to a manufacturing business; the salespeople, sales order clerks, warehouse managers, accounts receivable clerks, and others are spending most of their day supporting the process. If the process is efficiently designed and managed and the employees are properly trained, workers will have enough repetition to become efficient in their daily tasks.

Many business processes, however, are performed sporadically. The effectiveness of these processes can be poor, especially when they cross functional boundaries. Many times the work "falls through the cracks," not necessarily through negligence but due to a lack of repetition. For example, the process of establishing credit limits occurs occasionally, and requires coordination between Sales, which identifies new customers and gathers basic data (contact names, addresses, terms and conditions) and Accounts Receivable, which must evaluate the customer's credit history to establish a credit limit. Unless the process of establishing a credit limit is managed properly, a new customer's order may be blocked for an unacceptable length of time. For sporadic processes, a workflow tool can automate the process to ensure that the tasks are performed in a timely and correct manner.

Workflow tools are software programs that automate the execution of business processes and address all aspects of a process, including the process flow itself (the logical steps in the business process), the people involved (the organization), and the effects (the process information). ERP software provides a workflow management system that supports and speeds up business processes. It enables employees to carry out complex business processes and track the current status of a process at any time.

The SAP R/3 workflow tool integrates organizational data (which indicates how an authorized worker is supposed to perform a transaction) with business transactions. SAP's internal e-mail system lets you use workflow to view links between work and various transactions. These links, called **workflow tasks**, can include basic information, notes, and documents, as well as direct links to business transactions. The workflow tasks can be monitored by the SAP system, and if the tasks are not completed on time, the workflow system can automatically take various actions, including changing the workflow task priority and sending e-mail reminders to the employees responsible for the work.

Consider vacation requests. Employees performing critical operational tasks may be required to request time off from work in advance. This is a fairly irregular process that can create operational and/or employee morale problems if not handled properly. As such, it is a prime candidate for a workflow tool. In SAP, the Workflow Builder is used to define the process behind a workflow. In addition to defining the business process steps, the software identifies individuals involved in the process and sets other process parameters. Figure 7-14 shows a Workflow Builder screen that manages an employee's request for time off.

FIGURE 7-14   SAP R/3 Workflow Builder screen

Figure 7-15 shows the first step in this process, the absence request screen that the employee uses to request time off from work. Once the employee completes the screen requesting time off from work, the request becomes a workflow item in the Business Workplace for the employee's supervisor. The Business Workplace is a collection of SAP programs that provides a number of functions, including e-mail, a calendar, and workflow, similar to other comprehensive communication/collaboration software packages like Microsoft Outlook or IBM's Lotus Software.

FIGURE 7-15   Absence request screen

Figure 7-16 shows a request for time off in the workflow inbox of the manager's Business Workplace. From this screen, the manager can either approve or deny the request. If the request is approved, the employee is notified by e-mail that the request has been approved. If the request is denied, the manager can include a note explaining the rationale. The rejected request will be sent to the employee's workflow inbox, where that person can either modify the request or cancel it.

Workflow provides a number of useful features. Employees can track the progress of workflow tasks, reviewing their status at any time. Reminders can be sent to the employee(s) responsible for a task if it is not completed on time. Managers can build flexibility into workflow tasks regarding who can perform the task. For example, suppose that the Accounts Receivable (A/R) department has three employees who can set credit limits.

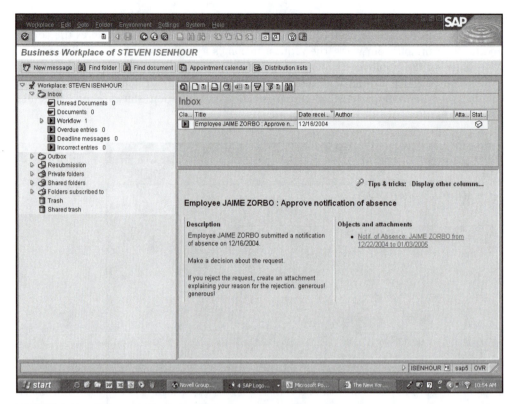

FIGURE 7-16   Manager's Business Workplace with workflow task

You could create a workflow task to set a new customer's credit limit and send it to the workflow inbox of all three A/R employees. When one of the employees completes the task, the system removes it from all three workflow inboxes. You can also generate statistics on the number of workflow tasks handled by each employee and the average time taken to complete the tasks, resulting in better staffing decisions.

For regular, day-to-day business processes, workflow tools are not required. But for sporadic processes that are repeated frequently enough to justify the development costs, workflow tools are a powerful way to improve process efficiency and effectiveness.

IMPLEMENTING ERP SYSTEMS

You learned about some of the issues and challenges in implementing ERP in Chapter 2. Implementation was particularly challenging in the late 1990s as many firms rushed to implement ERP systems to avoid the Y2K problem. Because of the explosive growth in demand for ERP implementations at that time, there was a significant shortage of experienced consultants. Since 2000, the pace of implementations has slowed significantly. Most Fortune 500 firms have implemented an ERP system. The current growth area in ERP implementations is in the small to midsize business market, and vendors have been developing products, including Microsoft Great Plains and SAP Business One, tailored to this market.

Recall that the implementation of ERP is an ongoing process, not a one-time task. ERP systems are extremely complicated and no company takes full advantage of all of the capabilities in the system. Firms that implemented ERP systems to avoid the Y2K problem likely installed ERP systems that only covered the basic functionality necessary to operate the business through the Y2K transition. Many of these firms are now looking to leverage their ERP systems to improve their business processes, which means that these companies are involved in implementation projects that, while smaller than the initial ERP implementation project, still require effective management.

ERP System Costs and Benefits

As you learned in Chapter 2, ERP implementation is expensive (usually ranging between $10 million and $500 million, depending on company size). Among its costs are:

- Software licensing fees: ERP software is quite expensive, and most ERP vendors charge annual license fees based on the number of users.
- Consulting fees: ERP implementations require the use of consultants with detailed knowledge of how to configure the software to support the company's business processes. Good consultants have extensive experience in how ERP systems function in practice, and they can help companies make decisions that avoid excessive data input while capturing the needed information to make managerial decisions.
- Project team member time: ERP projects require key people in the company to guide the implementation. These team members have detailed knowledge of the company's business and work with the consultants to make sure that the configuration of the ERP software will support the company's needs. This means that these workers are frequently removed from their daily responsibilities to work on the implementation project.
- Employee training: Project team members need training in the ERP software so that they can work successfully with the consultants in the implementation. The team members frequently work with training consultants to develop and deliver company-specific training programs for all employees.
- Productivity losses: No matter how smooth the ERP implementation, companies normally lose productivity during the first weeks and months after switching to the new ERP system.

Companies must identify a significant financial benefit that will be generated by the ERP system to justify the money they will spend on it. The only way companies can save money with ERP systems is by using them to support more efficient and effective business processes. This means that implementation projects should not recreate the company's current processes and information system in a new ERP package, which is a very real possibility. SAP provides all source code with its R/3 package, which means that the user can see how its ERP system is designed and can alter the package through its internal programming language, called Advanced Business Application Programming (ABAP), which you first learned about in Chapter 2. With access to the SAP R/3 source code, it is possible for a company to spend a significant sum of money on software code development to avoid changing its business process to the best practices designed into the ERP software, instead changing the software to fit its own existing, outdated methods.

Finally, companies must manage the transfer of data from their old computer systems to the new ERP system. In addition to managing master data such as materials data, customer data, vendor data, and so on, a company must also transfer transaction data, which includes sales orders and purchase orders, many of which are in various stages of processing—a challenging task.

IMPLEMENTATION AND CHANGE MANAGEMENT

How does a company make sure that its ERP investment pays off in increased profitability? The key challenge is not in managing technology, but in managing people. An ERP system changes the way people work, and for the system to be effective, the change may be dramatic, going beyond the look of the software and into how employees perform their tasks. Furthermore, business processes that are more effective require fewer people. Some employees will no longer be needed. It is no small thing to ask people to participate in a process that may not only change their day-to-day activities, but might also eliminate their current jobs.

Managing the human behavior aspects of organizational change is called **organizational change management (OCM)**. Do not underestimate the importance of this part of the implementation process. One of the keys to managing OCM is to realize that people do not mind change, they mind *being* changed.[3] If the ERP implementation is a project that is being forced on the employees, they will resist it. If it is viewed as a chance to make the company more efficient and effective by improving business processes, and if these process improvements will make the company more profitable and therefore provide more job security, then there is a greater likelihood that employees will support the implementation efforts. As the previous section mentioned, the best way to improve a business process is to have the people most familiar with the process use their experience and creativity to develop process improvement ideas. When employees have contributed to a process change, they have a sense of ownership and will likely support the change.

Implementation Tools

Many tools are available to help manage implementation projects. Process mapping, described previously, is perhaps the most critical. For an ERP implementation to go smoothly and provide value, it is critical that a company understand both its *current* processes and the state of the process *after* implementation.

SAP provides Solution Manager, a tool that helps companies manage the implementation of SAP R/3. In Solution Manager, the R/3 implementation project is presented in a five-phase Implementation Roadmap. The five phases are:

- Project Preparation (15 to 20 days)
- Business Blueprint (25 to 40 days)
- Realization (55 to 80 days)
- Final Preparation (35 to 55 days)
- Go Live and Support (20 to 24 days)

Figure 7-17 shows an example of this roadmap. The left side of the Solution Manager screen shows a hierarchical menu structure that organizes each step in the process, and on the right side of the screen are the detailed items and explanations for one step.

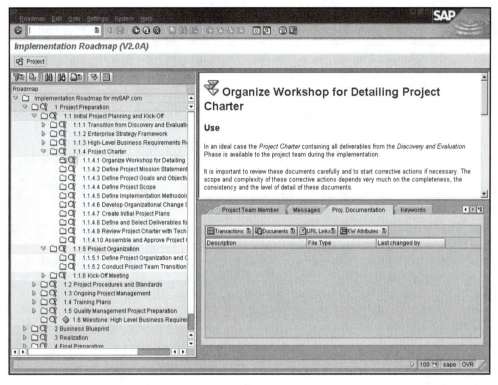

FIGURE 7-17   Implementation Roadmap in Solution Manager

Solution Manager has tools to support each phase in the roadmap, including documents, reports, white papers, and planning tools. The first phase of the Implementation Roadmap is Project Preparation. Some of the tasks in Project Preparation include organizing the technical team, defining the system landscape (including servers and network), selecting the hardware and database vendors, and, most importantly, defining the project's scope—what the project is to accomplish. A common problem in ERP implementations is **scope creep**, which is the unplanned expansion of the project's goals and objectives. Scope creep causes the project to go over time and over budget and increases the risk of an unsuccessful implementation. Defining the project's scope ahead of time helps prevent this problem.

The second phase, the Business Blueprint, produces detailed documentation of the business process requirements of the company. The Business Blueprint provides a detailed description of how the company intends to run its business with the SAP R/3 system. Process mapping is critical in the Business Blueprint phase. The Business Blueprint guides consultants and project team members in configuring the SAP R/3 system (which occurs in the third phase). During the Business Blueprint phase, technical team members determine the method of data transfer from the firm's existing computer systems (called legacy systems) that will either be replaced by the R/3 system or will continue to function with the R/3 system through an interface.

In the Realization phase (the third phase), the project team members work with consultants to configure the R/3 software in the development system. The team also develops any necessary ABAP code or other tools (such as third-party software packages) and creates the required connections to the legacy systems.

The fourth phase, Final Preparation, is critical to the success of the implementation project. Tasks in this phase include:

- Testing the system throughput for critical business processes (determining whether it can handle the volume of transactions)
- Setting up the help desk where end-users can get support
- Setting up operation of the Production System and transferring data from legacy systems
- Conducting end-user training
- Setting the Go Live date

When scope creep occurs in a project, it is commonly not discovered until well into the Realization phase, when the team begins to miss deadlines and the costs begin to exceed the budget. By the time the scope creep is discovered and its impact is understood, there is little management can do to correct the problem, as most of the time and budget have been spent. Management can choose to shorten or omit the Final Preparation phase, which means that testing of the system and training of employees is reduced or eliminated. Unfortunately, with reduced testing, errors in configuring the system are not discovered until it is put into use. Likewise, with reduced training, employees do not know how to use the system properly, which can create a complicated chain of problems due to the integrated nature of the system. Any cost savings gained by shortening the Final Preparation phase are overshadowed by productivity losses and consulting fees in the Go Live phase.

In the fifth and final phase, Go Live, the company begins using the new ERP system. Wise managers try to schedule the Go Live date for a period when the company is least busy. Setting up a properly staffed help desk is critical for the success of the Go Live phase, because users have the most questions during the first few weeks of operating with the new system. The SAP R/3 project team members and consultants should be scheduled to work the help desk during the first few weeks of the Go Live period. Although significant testing of the system and settings should have been done throughout the project, it is not possible to test all the settings and thoroughly evaluate the throughput of the system. Therefore, monitoring of the system is critical so that changes can be made quickly if the performance of the system is not satisfactory. Finally, it is important to set a date at which point the project is complete. Any enhancements or extensions to the system should be managed as separate projects, not as extensions of the original implementation project.

System Landscape Concept

SAP recommends a system landscape for implementation like the one shown in Figure 7-18. In this system landscape, there are three completely separate SAP systems, designated as **Development (DEV)**, **Quality Assurance (QAS)**, and **Production (PROD)**. The development (DEV) system is used to develop configuration settings for the system as well as special enhancements using ABAP code. These changes are automatically recorded in the **transport directory**, which is a special data file location on the DEV server. These changes are imported into the QAS system, where they are tested to make sure that they function properly. If any corrections are needed, they are made in the DEV system and transported to the QAS system. Once the configuration settings and ABAP programs pass testing in the QAS system, all settings, programs, and changes are transported to the PROD system, the system that the company uses to run its business processes.

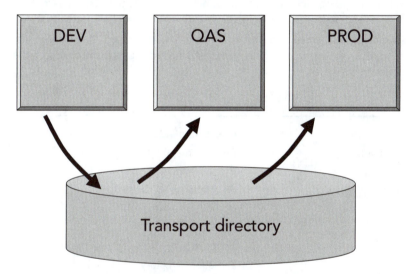

FIGURE 7-18   System landscape for SAP R/3 implementation

The use of separate systems is important during the initial implementation of an SAP R/3 system, and it is even more important after the Go Live phase. All software packages have occasional updates, and having systems available to test these updates before applying them to a production system can prevent problems. If a company wishes to use features of the SAP R/3 system that were not included in the initial project implementation, then the company should have a process like the one SAP provides to manage changes to the production system in a controlled fashion.

Chapter Summary

1. The concept of business processes has been an important theme throughout this book. ERP systems are designed to provide the information, analysis tools, and communication abilities to support efficient and effective business processes. This chapter introduced process modeling as a fundamental tool in understanding and analyzing business processes.

2. Process mapping is one process modeling tool that uses graphical symbols to document business processes. Other methodologies include hierarchical modeling, deployment flow-charting, event process chain diagramming, value analysis, and business process improvement. SAP's Solution Manager, a set of tools and information that can be used to guide an implementation project, is included in SAP R/3 to help manage the implementation of the ERP software.

3. SAP's System Landscape was introduced to show how changes to the ERP system during implementation (and beyond) are managed.

4. The critical issue of organizational change management was discussed. Most challenges to ERP implementation involve managing personnel and their reactions to the change, rather than managing technical issues.

Key Terms

Business process innovation (BPI)	Process boundary
Business process reengineering	Process mapping
Continuous improvement	Process model
Deployment flowcharting (swimlane flowcharting)	Production (PROD) system
	Quality Assurance (QAS) system
Development (DEV) system	Quality circles
Dynamic process modeling	Scope creep
Event process chain (EPC)	Transport directory
Flowchart	Value analysis
Gap analysis	Workflow
Hierarchical modeling	Workflow tasks
Organizational change management (OCM)	

Exercises

1. Develop a process map for the process a professor must follow to prepare a multimedia presentation of a concept for students. Developing a multimedia presentation can require various steps:

   • First, if the concept is not abstract, then it may be possible to record the subject.

   • If you have access to the subject of your multimedia presentation, then you should record audio, video, or graphic images as appropriate.

   • If your concept is abstract or you do not have access to it, then you should search to see if someone has already created media that illustrates the concept sufficiently for your use.

- If an alternate source does not exist, then you must create drawings, animations, or simulations of the concept.

- If you find the existing material, then you have to obtain copyright clearance unless it is public domain material or your use is covered by the Fair Use Guidelines (check with your instructor on this).

- If the media are not in digital form, then you must digitize it before distribution to students.

2. Develop a process map for the process of handling food shipments at a FoodMore Grocery Store:

- If a truck that contains perishable items (such as dairy, meat, or produce) arrives at a FoodMore store, the temperature of the truck has to be checked to ensure that the items were transported properly. Each product category has a product data sheet that defines the acceptable temperature range for the truck. If the truck is not in the proper temperature range for the product category, then all items of that category are rejected. More than one product category may be on a truck if the temperature ranges for the products are compatible.

- As cases are removed from the truck, they are checked for damage. Damaged cases are rejected. The expiration dates are checked against the date information on the product data sheets to make sure each case will have an acceptable minimum shelf life. If a case has an unacceptable expiration date, then the case is rejected.

- Differing processing steps are followed for all accepted cases depending on whether they are perishable. Once a case of perishable food is accepted, it is moved immediately to the cooler to wait further processing. Some products, like beef, pork, whole chicken, and cheese, are bought in bulk and must be cut and packaged before being put on the shelves. The processing procedures are defined in the product data sheets for each food. Bulk foods are processed on an as-needed basis, and a stock clerk takes the processed packages to the appropriate location in the store immediately after they are processed and packaged. The stock clerk also provides one last check of package integrity before placing the items on the shelves and display cases. If the package is damaged, the item is discarded.

- If an item is not perishable, it is placed in the storeroom and then moved onto a store shelf when the number of items on the shelves gets low. The stock clerks also check the package integrity of the nonperishable items before placing them on the shelves, looking for things like dents in cans. Depending on the level of damage, dented cans may be marked down in price and placed on the shelves in a special reduced price area. The clerks might discard the items if the damage is too great.

- Once items are on the store shelves, they must be monitored periodically to make sure that the expiration dates have not been exceeded. If an item is getting close to its expiration date, the price may be marked down for quick sale. If the expiration date has been exceeded, it must be discarded.

3. Develop a process map for following process, which describes the sales process for Daisy Scout cookies. The cookies generate income for individual scout troops and for the regional scout council. Use a deployment (swimlane) format. Can you suggest any changes to improve the process?

- The Daisy Scout troop leader for each troop receives the order forms from the local Daisy Scout council office. The leader passes out the order forms to the girls at the beginning of the cookie campaign. The girls go out and book cookie orders during the first order period. If a scout completes the entire form, she can turn it in to the troop leader and get another form. At the end of the first order period, the scout leader collects all the cookie order forms, makes copies, and gives the forms to the "cookie mom" (a volunteer who manages the cookie orders for the scouts). The cookie mom tabulates all of the orders and then delivers a consolidated order form to the Daisy Scout council cookie coordinator. The troop order must be for full cases (12 boxes per case). The troop's orders might not total full cases, meaning that it will receive more cookies than the scouts sold.

- When the cookie manufacturer delivers the cookies to the Daisy Scout council office, the cookie mom is contacted and picks up the cookie order from the council office, checking the delivered quantity against the order. The cookie mom takes the order home and then breaks the cases down into groups for the individual orders. She delivers these orders to the girls at their next meeting. The girls deliver the cookies to the customers, collecting $3 for each box. The girls return the money to the Daisy Scout troop leader, who passes it on to the cookie mom. The cookie mom deposits a portion of the money in the troop's account and deposits the council's portion in the council bank account. This process must be documented at the bank and all receipts turned in to the council before more cookies can be ordered. The portion of the sales that the troop keeps depends on the sales level. Normally they collect $0.50 per box, but if the troop sells an average of 100 boxes per Daisy Scout, they collect $0.55 per box. The council also gets a percentage of the proceeds to support their operations. The troop can place a second order with the council coordinator if they deposit by the required date at least 50 percent of the money they collected from their first sales effort.

- The second order also has to be for case quantities, and the excess boxes from the first order must be considered. To avoid buying an entire case of cookies to get one or two boxes needed for the orders, the cookie mom can deal with cookie moms from other troops to buy excess boxes of cookies (although this practice is discouraged by the council office). When buying a box of cookies from another cookie mom, the moms have to agree on which troop gets to keep the profits. This cookie dealing can be reduced if the troop decides to have a direct sales effort. For example, they might set up a table at the local grocery store and sell boxes of cookies for cash. If the troop wants to do this, they have to get approval from the council office. The council office assigns the selling locations, so the troops should do this early in the process to get their choice of location. If that is the case, then the cookie mom can dispose of any excess boxes from her own troop's orders, although she may choose to help out other cookie moms by selling boxes to cover some of their orders (again, the sales price has to be determined).

- Once all the money from the second round of orders and the direct sales is collected, the money needs to be deposited in the troop and council accounts. All paperwork must be submitted to the council cookie coordinator by a given date. It is possible that the troop sold less than 100 boxes per girl in their first effort but achieved that goal with the second order and direct selling. In that case, the troop gets to keep $0.55 for all boxes sold, and the deposit needs to be calculated accordingly. Any leftover boxes can be used for snacks at the troop meetings, and are paid for with troop funds.

4. For the accounts payable process described below:

   a. Develop a process map.

   b. Analyze the value added by each step in the process. Determine whether the step adds real value, business value, or no value.

   c. Develop a process map for an improved process.

   - The accounts payable process at Frizz Master Hair Product Manufacturer begins when specific departments ask to purchase certain products or equipment. The department manager then approves the requests so that the buyers in the Purchasing department may purchase the requested materials or products.

   - Once the necessary product has been bought and the invoice has been generated, an invoice for payment is sent to the Accounts Payable (A/P) department. The A/P tasks are divided between two people. One person handles invoices for companies with names that start with the letters A through M, while the other person handles invoices for companies with names that begin with N through Z. As each A/P worker opens the invoices, they sort them according to the department responsible for the payment. All invoices that involve overhead (and therefore do not get directed to a specific department), such as utility and shipping invoices, are coded by the A/P clerk and entered into the accounting system for payment. All other invoices are sent by intercompany mail to their respective departments. Invoices for amounts under $1,000 are directed to the person who requested the purchase, while invoices over $1,000 are sent directly to the manager in charge of the specific department where the request originated. Once the originator or manager approves the invoice, it is sent back to the A/P clerk, who enters it into the accounting system.

   - The A/P clerk selects all approved invoices that have been entered into the accounting system and prints them as a list for the Accounting Manager, who marks the list to show which invoices are to be paid during the current check cycle. The Accounting Manager gives the annotated invoice list back to the A/P clerk, who reports to the Staff Accountant to retrieve a special disk that allows access to the check printing system. The disk controls access to the check printing system, and must be kept under lock and key, which is the responsibility of the Staff Accountant. The A/P clerk uses the disk to connect to the check printing system and print checks to be immediately mailed to the vendors, along with the corresponding invoice. The A/P clerk must keep written records of the check numbers and the amount of each check that is printed and mailed. Finally, the Account Manager double checks for errors and signs off on the check number and dollar amounts of each check after each check run.

5. Develop an EPC diagram for the following Human Resources process:

- The current recruitment process for Yellow Brook Photography takes approximately 90 days. It begins when a manager has a need for a position. The manager must complete a requisition and send it to the Human Resources (HR) department. HR reviews and assigns a number to the requisition and returns it to the manager for approval. He approves it, obtains the appropriate signatures, and then returns it to HR.

- Next, HR creates a job posting and announces the position internally first through the company's Intranet, bulletin boards, or a binder of current job openings. They collect responses for eight days. HR also solicits resumes from external sources by advertising. HR prescreens resumes and forwards data on qualified candidates to the hiring manager for review. The hiring manager notifies HR of candidates to interview. She also conducts phone screens; if the phone screen is promising, HR coordinates and schedules an on-site interview. Candidates interview with the hiring manager and with HR. HR records the interviews in an applicant flow log.

- Once a candidate is selected for hire, HR and the hiring manager prepare an offer and the background check is initiated. The hiring manager then must approve the offer and obtain the required signatures on an internal associate data/change form. Subsequently, she must extend the offer verbally to the candidate while HR sends the written offer, including a start date for work. Once the applicant accepts the offer, a drug screening is scheduled with the candidate, who must also sign and return the offer letter to HR. HR notifies the hiring manager of the candidate's acceptance. Finally, if the drug test comes back negative, the new employee completes "new-hire" orientation on the date hired.

End Notes

[1] Ronald M. Cordes, "Flowcharting: An Essential Tool," http://www.qualitydigest.com (January 1998).

[2] H. James Harrington, *Business Process Improvement: The Breakthrough Strategy for Total Quality, Productivity, and Competitiveness*, McGraw-Hill (April 1991).

[3] Michael Hammer and James Champy, *Reengineering the Corporation: A Manifesto for Business Revolution*, HarperCollins, 1993.

CHAPTER 8

ERP AND ELECTRONIC COMMERCE

INTRODUCTION

This chapter examines connectivity topics relevant to Enterprise Resource Planning (ERP) systems. As you have read, an ERP system lets a company accomplish things that cannot be done well, if at all, without such a system. In this chapter, you'll learn how effectively competing in high-volume e-commerce may be impossible without the infrastructure provided by ERP. In addition, you will learn how companies can integrate ERP systems with the Internet and "rent" ERP software from special-purpose software companies. You'll also learn about the basic components of Web services, with a focus on NetWeaver, SAP's Web services platform. You will also find out why XML is becoming the new markup language of the

Internet and how ERP companies are using it. Finally, you'll read about radio frequency identification devices (RFIDs) and their significance to managing the movement of goods in the supply chain.

ELECTRONIC COMMERCE BACKGROUND

Today most companies are conducting at least part of their business operations through electronic commerce, more commonly known as e-commerce. As you learned in Chapter 2, e-commerce is the conduct of business over the Internet. When people think of e-commerce, they often think of retail e-commerce, typified by companies like Amazon.com. Most of the business growth on the Internet, however, has been in the area of **business-to-business (B2B)** e-commerce, rather than in retail **business-to-consumer (B2C)** e-commerce. Let's look at B2B e-commerce.

Business-to-Business E-Commerce

Business-to-business e-commerce is defined as buying and selling between two companies over the Internet. The companies might be manufacturers, suppliers, wholesalers, or retailers. B2B e-commerce sales are expected to approach $1 trillion in Europe alone by 2006.[1]

Business-to-business e-commerce is transforming the way companies work with each other. For example, industry-specific auctions allow companies to buy raw materials at the lowest possible cost. Say a food company's purchasing manager wants to buy honey. She can go to an auction site—either for the food industry or a general "wholesale/retail only" auction site—and get competitive bids simply by posting the request rather than requesting competitive bids from each supplier independently by e-mail or fax.

Electronic Data Interchange (EDI)

B2B e-commerce is not new. Prior to the development of the Internet, companies electronically transferred purchase orders through a system known as **electronic data interchange (EDI)**. EDI is an electronic computer-to-computer transfer of standard business documents. Companies have been doing EDI since the 1960s and originally used telephone lines to transfer data.

Companies can set up their own private EDI networks to communicate directly with their suppliers' systems. EDI networks are very expensive, so many companies subscribe to a **value-added network (VAN)**, an intermediary Internet-based network run by an outside EDI service provider. With EDI, when a company needs to order a raw material, it can place the order electronically. EDI should not be confused with a simple e-mail message between two parties that reads, "Dear Joe, please send us 500 gallons of honey." In EDI, the message is sent on a standardized business transaction form, following a specific computer protocol.

The benefits of EDI have been enormous:

- Costs of paper, printing, and postage have almost disappeared from ordering systems.
- Errors have been minimized because orders are not manually entered into the supplier's information system.

- Ordering is fast and efficient. As a result, large companies can force their smaller suppliers to use EDI, through a VAN if necessary. The supplier pays the VAN's fees based on the amount of information sent and received.
- Suppliers and buyers are "locked" into business relationships: Once a company sets up an EDI system with its supplier, changing suppliers becomes a major inconvenience. Most buyers don't change suppliers, hence locking in the relationship. This is an advantage for suppliers and buyers as long as both remain satisfied.

Internet-Based Procurement

Even though EDI has been useful, companies are moving from EDI to **Internet-based procurement**, which is the use of Internet technologies for procurement activities (recall that procurement is the buying of raw materials for manufacture or finished goods to resell). Internet-based procurement provides the following benefits:

- It is less expensive to use the Internet than private EDI networks.
- Purchasing costs are further reduced as suppliers compete for orders on the buyer's Web site.

Locking in suppliers often does not occur in Internet-based procurement. Buyers tend to ignore suppliers who cannot compete on price, instead focusing on relationships with viable suppliers.

Buying and selling goods on the Internet has propelled the concept of electronic marketplaces. To understand this concept, think about how a farmer's market works. Farmers, growers, and gardeners bring their produce to a central location in town, where customers can compare sellers' goods on quality and cost. It is efficient for both buyers and sellers. Being grouped together, sellers gain access to more customers than they would on their own. Customers can shop for a variety of items in one place, making it easier to find the best quality at the lowest cost.

An electronic marketplace is a gathering place for buyers and sellers on the Internet. It can be run with or without a central operator. Marketplaces without central operators are simply lists of Web sites directing buyers to certain products. Marketplaces with central operators facilitate and expedite the buying and selling of goods. Businesses using this type of marketplace may have to pay fees to join, and usually pay fees to use the marketplace.[2] **Exchanges** are one type of B2B electronic marketplace. Exchanges typically focus on a single industry.

In 2000, Ford Motor Company, General Motors, and DaimlerChrysler announced that they were putting together a B2B automotive trading exchange called Covisint. This venture is managed in conjunction with Commerce One and Oracle, which is providing the software. Each company, along with thousands of suppliers, can buy and sell automotive parts on this Web site.[3] Purchasing auto parts at Covisint is fast and efficient, allowing auto manufacturers to maintain lower inventory levels. These savings could translate to more than 10 percent of the purchase price of a car.

Chemical buyers, sellers, and traders can find their products at chemconnect.com. The Web site claims the following advantages:

- Buyers can find the best prices without traditional negotiations.
- Faster contracts are completed between buyers and sellers.

- Buyers and sellers can gain access to new worldwide markets and new trading partners.
- Instant market information is available to all parties.

ChemConnect members use the marketplace exchange to react more quickly to changes in the often volatile chemical market than they could using more traditional methods of buying and selling. For example, Vanguard Petroleum Corporation places bids and offers simultaneously on ChemConnect to multiple customers. About 15 percent of Vanguard's spot purchases and sales of natural-gas liquids are conducted on the ChemConnect site. Vanguard can now publicize its offers to 150 potential partners in only a few minutes; prior to its involvement with the exchange, it took an entire day to get the information out to that many companies.[4]

Another type of industry marketplace is the private exchange, where membership is restricted to select participants. Siemens hopes to save $900 million per year in using an exchange limited to approved suppliers. Private exchanges can be attractive to businesses because they are limited to trading partners with which the responsible party has an established relationship. For example, IBM runs a private exchange for its 25,000 customers and suppliers; IBM's competitors are prohibited from participating in the exchange.[5] Likewise, Volkswagen opened a private exchange in 2003 after deciding that Covisint did not promise some of the capabilities it was looking for in an exchange. Since it opened the site, Volkswagen has bought $5.2 billion in goods on its private exchange, slashing procurement costs by up to half. It has cut contract negotiations from three months to as little as a day, reducing costs by cutting weeks out of project cycles.[6] Jupiter Research estimates that by the end of 2005, one-third of all businesses with revenue of more than $1 billion will operate private exchanges.

Internet Auctions and Reverse Auctions

As previously mentioned, B2B e-commerce allows companies to do online bidding through auctions and reverse auctions. **Reverse auctions** feature one buyer and many sellers. A company can use a standard auction to put up for bid its products or even its obsolete equipment and at the same time use reverse auctions to request bids from suppliers for goods or services.

For example, Fitter Snacker needs raw materials such as oats and wheat germ, which for FS are essentially commodities. **Commodities** are items that are widely available at a standard level of quality; the only thing that varies between one commodity and another is price. The company can go to a "bidding" Web site—one that seeks bids from sellers rather than buyers—and set up a program to run overnight. The program uses the Internet to solicit bids to supply those raw materials. In the morning, FS's purchasing agent can choose the lowest price offered for the oats and wheat germ.

Internet-based auctions are changing the way in which commodities are purchased. A few years ago, commodities would have been purchased through a supplier, or intermediary, who negotiated prices of raw materials with suppliers. Now, the Internet and its bidding programs have threatened the intermediary's role and made the buying process more efficient. In a sense, the Internet has replaced the intermediary. Pricing is open and dynamic, meaning that competitors can see each others' bids (although they probably can't see who is making the bids). Epsilon Products Company, a producer of polypropylene, has used a reverse auction on chemconnect.com to reduce its cost by 5 percent.[7]

Dynamic pricing is not only forcing out intermediaries, it is also putting pressure on sellers to be flexible. This requires a seller's accounting and logistics operations to be in excellent shape (requiring an ERP system) before trying to sell in the auction market, as competition is based on price.

Electronic Commerce Security

E-commerce brings with it a major concern: security. No company (or individual, or government agency, for that matter) is immune to security breaches. In the past, large firms have been shut down due to various types of systems attacks, including denial-of-service attacks. In **denial-of-service (DoS)** attacks, attackers block access to a Web-based service through a variety of means, including bombarding the sites with so many messages that the site can't handle the volume. In February 2000, Buy.com, Amazon.com, CNN.com, eBay.com, E*Trade.com, ZDNet.com, and Yahoo.com were all temporarily knocked out because of DoS attacks. Yahoo.com lost approximately $500,000 due to the three-hour attack.[8] Security is an ongoing effort. E-commerce companies use virus-scanning software, encryption, intrusion detection, and other measures to protect their networks, their Web sites, and the privacy of customer data.[9]

E-COMMERCE AND ERP

You might ask yourself, what does e-commerce have to do with ERP? The answer is that each technology complements the other, and each is necessary for success. An offline company cannot compete with companies offering similar goods and services if the competitors also provide the added convenience of doing business over the Internet. Without ERP, a company cannot fill orders—either Web orders or traditional orders—expeditiously. Here's why.

When a company receives an order through its Web site, the company should not merely file or print orders for later handling. The orders should be efficiently fed into the company's marketing, manufacturing, shipping, and accounting systems—a series of steps sometimes called **back-office processing**.

An efficient back-office operation is crucial for any company's success. E-commerce often exacerbates problems and reveals weaknesses in current back-office systems. Amazon.com invests its profits back into warehouses and other support that the company needs to keep its back office in order, supporting its continued success. Some Web-only businesses are worried by the entry of massive retailers like Wal-Mart into their Web marketplaces. Large companies already have well-established, integrated back-office and distribution systems, which are typically more expensive to develop than e-commerce systems, giving the large retailers speed-to-market and financial advantages.

Some companies with unintegrated information systems have built Web sites before creating an integrated back-office system. As a result, those companies often can't fill orders in a timely fashion. Many e-commerce businesses suffered from a lack of integration during the 1999 holiday season. The online toy retailer eToys.com announced less than a week before Christmas that it would *not* be able to fill all its Web orders. Surprisingly enough, all the toys were in the warehouse, but the company couldn't organize basic warehouse functions such as picking, packing, and shipping to get toys to consumers on time. Integrating the Internet front-office operation and the ERP back-office operation is fundamental in today's business environment.

Fitter Snacker and E-Commerce

Fitter Snacker has traditionally sold its NRG bars through the sales force in either the Direct Division or the Distributor Division. Now FS's customers want to order directly from a Web site. Currently, Fitter Snacker has neither a Web-ordering system nor an ERP system. In the current annual-planning cycle, FS executives are looking at two IS investment options: (1) implement a Web-based ordering system, or (2) implement an ERP package. Company executives would like to make both investments, but they are reluctant to make two large investments in one year because of the cost and disruptive nature of the implementation efforts. While implementing an ERP system would improve its business processes, the company is experiencing significant customer pressure for Web-site ordering.

Suppose FS executives succumb to customer pressure and implement a secure Web site, allowing customers to place orders online. FS does not know exactly what percentage of customers would use the system (although if FS already had an ERP, that data would be readily available). To promote sales and be competitive with other online merchants, the company promises delivery from Web site orders in five working days, which seems reasonable, based on FS's past performance.

After the Web site's implementation, orders come in more frequently (which is good; it means that business is thriving). However, orders now have shorter lead-times because without an ERP system, Internet orders are saved to a text file, printed, and then manually transferred into the order-entry system described in Chapter 3. The order is then handled using the warehouse procedures discussed in Chapters 3 and 4. The increase in orders is putting a strain on the company's back-office operations.

A customer who orders NRG bars from the Web site *thinks* he will be receiving the bars within five business days, as stated on the company's Web site. After 24 hours have elapsed, however, the customer receives an e-mail advising him that the NRG bars are out of stock and will not be available for at least one week, maybe two. The company's unintegrated information system, barely adequate for personal and phone-based ordering, is simply too cumbersome to handle the Internet orders. With e-commerce, customers have increased service expectations, and FS's unintegrated system cannot meet those expectations.

Now consider an alternate scenario. If Fitter Snacker implements an ERP system first, and later connects it to a Web-ordering system, Internet orders could flow directly into the ERP system. As customers place orders through the Web system, the ERP system could provide accurate delivery date quotes. With this information, customers could make an informed choice and would be more satisfied. In addition, accurate and timely order data would support more effective production planning. And of course this discussion doesn't begin to address all the other ERP benefits that FS would enjoy.

Recent studies on back-office systems concluded that an attractive Web site does not provide enough benefit on its own for an e-commerce business to stay afloat. The conventional back-office systems must be in place and operating correctly for the business to flourish. As with any kind of business, effective infrastructure is key for e-commerce success.

Exercise 8.1

a. Assume you are opening a small business to sell a product that interests you. Some ideas might be clothing for your favorite sport or some new toys for your dog. Also assume that you will only have an Internet presence with this business—no physical store. What do you think is involved with setting up an e-commerce business? Use the Internet or interviews with people you know who have tried this. List the steps involved. Report your findings to the class.

b. Do you think customers' expectations are different when ordering on the Web versus ordering in a traditional store? Use some of your own experiences to answer this question.

c. Assume Fitter Snacker implemented both an ERP system and a Web ordering system. Develop a process map of the steps needed to fill a Web order. See Chapter 7 for details on process mapping. Be sure to include a description of how the ERP system interfaces with the Web page.

d. Develop a plan for an e-commerce business that sells concert gear such as t-shirts, hats, and pins. How will it establish a competitive advantage? What role will back-office systems play?

USING ERP THROUGH AN APPLICATION SERVICE PROVIDER

Many companies today outsource some of their operations to an outside service provider, sometimes called a third party. In the next section, we will look at how outsourcing can simplify the management of ERP systems.

Application Service Providers

An **application service provider (ASP)** is a company that provides management of applications for a company over a network. Usually that network is the Internet. Companies can outsource their e-mail software, accounting programs, or other programs to an ASP.

The company using the ASP does not own the software or the hardware and does not employ the people running the outsourced applications. The ASP owns the hardware and the rights to the software; it also employs the workers who administer the software. The users of the system, of course, are the company's employees. Suppose an ASP were to run an ERP package such as SAP R/3 for Fitter Snacker. The ASP would own the rights to use the SAP software. The ASP would also own the server on which the SAP software runs. The ASP could also provide consulting services for configuring the ERP system, so that Fitter Snacker would not have to find its own ERP specialists. The ERP program would be delivered to FS workers over a network, probably the Internet. The ASP would charge either a monthly or yearly fee, or it could charge a per-use fee for the system based on the number of users at FS. Thus, an ASP can provide ERP software with a much lower start-up cost, making it possible for smaller companies to use ERP systems when they otherwise could not afford the costs of installing and maintaining their own system. Figure 8-1 outlines the details of running an ERP system in-house versus with an ASP.

Fitter Snacker in-house ERP

Fitter Snacker ASP ERP

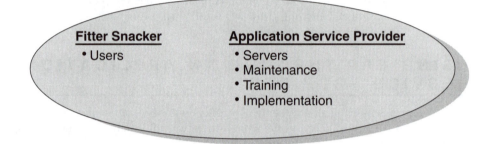

FIGURE 8-1 ERP responsibilities in-house versus with an ASP

Advantages of Using an ASP

Many companies find it advantageous to use an ASP for ERP and other information system applications. Some of the benefits include:

- *Affordability:* Companies that cannot afford their own ERP system can now "lease" one on a monthly basis, avoiding the high cost of obtaining the hardware and software and hiring and training support personnel. ASP services can be received through the Internet using either a Web browser or the ERP system's graphical user interface (GUI) software.

- *Shorter implementation times:* The time needed for ERP implementation is shorter for those who implement ERP through an ASP. An ASP should have experience with implementing and maintaining ERP software. If it does not, the company should find another ASP. They have servers, telecommunications, and highly trained personnel already in place.

- *Expertise:* ASPs are experts in delivering IS applications. They do all the maintenance, including execution of backups, training, and customizing of the system. Customers of ASPs do not need to hire additional IT personnel. ASPs can also run information systems more efficiently because they do it on a large scale. ASPs can spread fixed costs over many users, thus achieving economies of scale that might translate into a lower total cost of ownership. The availability of IT talent is tight in the current market; an ASP might have a better chance of attracting and retaining a talented workforce than a small manufacturing company like Fitter Snacker.

Disadvantages of Using an ASP

There are some potential problems with using an ASP:

- *Security:* Companies using ASPs are turning their information systems over to a third party. They must be confident that the ASP has a high level of security. How hardware will be shared is also a security concern. The ASP will have multiple users on a single piece of hardware. Each customer's data must be shielded from other customers. While these are valid concerns, ASPs often have better security than a small company can have on its own.
- *Bandwidth/response time:* The telecommunications channel from the ASP to its customers must be fast enough to handle multiple users. An ASP's servers must be sufficient in terms of processing capabilities.
- *Flexibility:* An ASP should be flexible in working with its users and satisfying their requests for processing modifications.
- *No frills:* An ASP can usually provide basic ERP systems well, but asking for unusual configurations may cause problems. An ASP might not allow for third-party add-ons, either. Further, the ASP might not want to do custom software development using the SAP R/3 programming language, ABAP.
- *Technical, not business:* An ASP knows the technical aspects of the software, but it will need the customer to define the business processes and make the configuration decisions.

ANOTHER LOOK

Using ERP through an Application Service Provider

SAP software is run at over 400 universities as part of SAP's University Alliance. The University Alliance program grants access to SAP's ERP software for classroom use. At the time the University of Delaware began its University Alliance program in the 1990s, each university was required to purchase its own server (which at the time could cost upwards of $50,000) and had to train its own system administrators. With over 100 universities in the Alliance, managing all of the systems became a significant problem for SAP. On the academic side, operating the system was a chore for the university's systems administration personnel because managing the SAP system was usually an added duty on top of their normal duties. To address these problems, SAP's University Alliance program developed its own ASPs, which it termed University Competency Centers (UCC). Rather than having every university administer its own system, the five universities that serve as UCCs and host the other members of the University Alliance, meaning that the UCCs act as ASPs for the other members. For SAP, the task of managing five UCCs is much easier than managing over 100 separate systems. The UCCs have worked together to develop their

continued

competency in administering systems for education—a specialized task. For students, accessing an SAP system from a UCC seems no different than accessing a system at their own university, except that the UCC system is usually much faster, and the software is current and ready for use.

Questions:

1. What are the advantages of the University of Delaware using an ASP for its SAP delivery? Are there any disadvantages?
2. What are the challenges to a university to running a complicated program like SAP? Can you relate those challenges to small companies?

Cirque du Soleil is another organization using an ASP for ERP delivery. This traveling performance company tracks 700 tons of costumes, personal belongings, furniture, kitchen equipment, and school and office supplies hauled in 350 trailers across the country, supporting performers in five concurrent tours and four permanent shows.[10] The Canadian company runs six SAP modules that are hosted by IBM, which provides a range of ASP services. From IBM, Cirque du Soleil receives 24-hour-a-day support, flexible bandwidth, storage, and extra server power at peak usage times.

The small to midsize business consulting firm AMI-Partners, Inc. predicts that the ASP market for small and midsize businesses (those with less than $1 billion in sales) is $3.8 billion in 2005, up from $2.9 billion in 2004.

Other Considerations

As with all forms of outsourcing, companies considering an ASP should carefully scrutinize the ASP's contract before signing. There may be hidden costs and potential problems.

ERP companies are excited about ASP capabilities because ERP vendors can provide ASP services too. Such an offering translates to extra profit and a steady income. Prior to delivering software through an ASP, ERP software companies relied on a small number of very large sales to make their yearly profit. Now income is steadier because ERP vendors can deliver and support software on a monthly basis. For example, an ERP vendor might have high revenues one quarter by securing a few, large sales, but may follow that with lower revenues the next quarter if no large software deals are finalized. By delivering software through an ASP, however, the lease payments provide a steady monthly income.

Exercise 8.2

Let's return to our example company, Fitter Snacker. Assume that under its newly appointed CIO, FS has made the decision to acquire an ERP system. Now the CIO must decide how to implement the system. After talking with the vendors of ERP software, the CIO realizes FS has two options for implementing ERP software:

a. Buy the rights to the ERP software and any new hardware required to run it, and also hire and train system administrators.

b. Run FS's ERP system off the Web through an ASP, which would deliver ERP services for a monthly fee.

Both types of implementations have advantages and disadvantages, as you have already seen. In this exercise, you must recommend one course of action to FS's CIO. For the first part of this exercise, write a memo to the CIO enumerating the pros and cons of each method. Use the table shown in Figure 8-2 to organize your thoughts.

Advantages of purchasing software and computers for ERP	Advantages of using an ASP to run ERP

FIGURE 8-2

After distributing your memo to various FS board members, the CIO needs to justify his decision. He would like a spreadsheet that analyzes the financial impact of both scenarios. Your job is to compare the monthly cost of using the ASP with the cost of an in-house ERP system. Weigh the pros and cons of each method, and then make a recommendation to the treasurer of FS. Keep in mind that Fitter Snacker's net income is $3.4 million (you can find more details in Chapter 5). Here are the details of the decision:

Option 1: Buying Computers and Software Rights

To set up its SAP system, FS must buy:

- *Database server:* The server would cost $70,000.
- *Application server:* FS needs a server to run the ERP application, which would cost about $40,000 since there aren't any servers available for use in the system. (ERP software is platform-independent; therefore, it can be run on different types of computers. Thus, a company can often use an existing server.)
- *PCs:* Some of FS's existing PCs could be loaded with the ERP software's graphical user interface (GUI) and be used to access the system. Because more FS employees will be connected to the new system, however, FS will need 10 additional computers. Total cost for the PCs would be $15,000.
- *Rights:* Rights to the ERP software for all users for five years would cost $500,000. The CIO does not know whether further outlays will be required after the fifth year, and therefore is limiting the analysis to the years 2005 through 2009.
- *Installation:* The ERP vendor will help install the system, but FS also needs to hire consultants for the six-month implementation. At $3,000 per day, the cost is estimated at $486,000.
- *User training:* With the purchase of the rights to the ERP software, FS employees receive training at a local training center. This training is for key personnel involved in the ERP implementation project. FS wants further training, however, for FS-specific business practices. The additional training will cost $2,000 per day for two weeks. This includes a training consultant to run classes at FS headquarters. With travel and lodging, the total cost is $23,000.

- *Ongoing consulting:* Once the system is up and running, FS will need to pay for consultants to help maintain the system. FS estimates that if it budgets for consultants to come in once a month for $3,000 a day, it should be able to have all employees' questions answered. The total yearly cost of additional consulting is $36,000.
- *Computer maintenance:* FS needs to make sure all PCs and servers run properly. To do this, FS would purchase a maintenance contract to cover all hardware. This contract costs about $1,000 a month, or $12,000 yearly.
- *Network and database administrator:* FS would need a full-time network and database administrator to run the system. Salary, including benefits, for a skilled person is $200,000 per year.

Option 2: Using ERP Through an Application Service Provider

The other option is to use an ASP to deliver ERP software. Estimated costs for this option are as follows:

- *PCs:* FS still estimates that it must purchase 10 more PCs because many users will now be accessing the computer system. Each PC costs $1,500, or a total of $15,000.
- *Maintenance on PCs:* The maintenance contract on all the PCs at FS costs $600 per month, or $7,200 yearly.
- *Software through the ASP:* The monthly cost of delivering ERP software to FS over the Web is $33,333, or $400,000 yearly.
- *Training:* Training of FS employees is provided by the ASP as part of the monthly software fee.

Make a recommendation to FS. Which option should FS choose—purchase the SAP software and computers outright, or use an ASP? Set up a spreadsheet that will add all the costs of each option. In each scenario, you must deal with the net present value (NPV) of money.

NPV is a way to figure out whether an investment is profitable, or in this case, compare outlay of funds from one method to another. NPV can be calculated over a number of years; in our case, we need a five-year outlay of funds for the ERP project. The syntax of NPV is =NPV (hurdle rate percentage, range of values) in an Excel spreadsheet. The values in the range can be positive or negative numbers. In our case, they are all outflows, but we can work with them as positive numbers. The hurdle rate is the rate of discount over the period. The hurdle rate is the minimum acceptable rate of return on a project that a company will accept.

Your spreadsheet should begin like the one shown in Figure 8-3 (with years continuing through 2009). Assume payments are made throughout the year as opposed to lump sums at the beginning of the year.

ERP Purchasing Options				2005	2006	2007	2008	2009
Option 1 - Buying computers and software outright								
Items				**2005**	**2006**	**2007**	**2008**	**2009**
Database server				70000				
Application server				40000				
10 PCs				15000				
Software - SAP				500000				
Consultant - initial (6 months)				486000				
Training (2 weeks)				23000				
Consultant - maintenance (1 day per month)					36000	36000	36000	36000
PC maintenance					12000	12000	12000	12000
Network administrator				200000	200000	200000	200000	200000
Total				1334000	248000	248000	248000	248000
NPV				$1,646,671.81				
Option 2 - Using an ASP								
PCs				15000				
PC maintenance					7200	7200	7200	7200
ASP cost				400000	400000	400000	400000	400000
Total				415000	407200	407200	407200	407200
NPV				$1,224,277.26				
Hurdle rate				20%				

FIGURE 8-3   Cost comparisons: buying versus renting

To complete this exercise, perform the steps that follow:

- Calculate the cost of the two methods of implementing SAP R/3 for five years. Use the spreadsheet as illustrated in Figure 8-3 as your guide. Use the NPV calculations to reference the hurdle rate at the bottom of the spreadsheet. Vary the hurdle rate, following the directions your instructor provides.
- Consider using different hurdle rates for each option. Why might that be applicable for this decision?
- Write a memo, with your spreadsheet attached, to the CIO. Answer this question: Which method should FS choose, and why? Be sure to consider both the qualitative aspects and the quantitative aspects of the choice. Also address the viability of the ASP.

Your instructor might assign the following additional exercises:

1. Use the Internet or library resources to research the use of ASPs. Find cases of companies that have been successful in using an ERP system through an ASP. Describe one success story in a memo to your instructor.
2. Would you give different advice to a smaller company than you would to a medium-sized company or to a large company regarding the use of ASPs? Why?

E-commerce is driving companies to connect their business applications, such as ERP, to the Internet to provide data sharing between companies. The combination of software tools that lets various programs within an organization communicate with other applications is called **Web services**. Not surprisingly, ERP companies are trying to stake out their territory in this new and lucrative field. SAP has invested a lot of time, money, and energy into its Web services platform, NetWeaver. NetWeaver, like other Web services products, allows various vendor applications to share data.

Companies are using Web services to work together in new ways. Dollar Rent-a-Car provides partners such as Southwest Airlines access to its internal system using Web services. This connection generates thousands of new auto rentals per year. FedEx's package tracking system is built on Web services. The technology has reduced FedEx's cost per inquiry from $2.14 to $0.04. Travelers Insurance Company has been able to cut its auto glass claim processing by 30 percent thanks to Web services.[11]

NetWeaver Tools and Capabilities

SAP's NetWeaver is a collection of components that support business transactions over the Internet. Included are modules named Enterprise Portal, Mobile Infrastructure, Business Intelligence, Master Data Management, and Exchange Infrastructure.[12]

The Enterprise Portal also goes by the name mySAP.com. It gives users complete access, or a portal, to all their work on a single screen, using links to all major aspects of their jobs. A **portal** is a customizable Web site that serves as a home base from which users navigate the Web. The Enterprise Portal acts as a central access point to a company's intranet, operating through a secure link on a browser. For example, a user in the Finance department could set up Enterprise Portal with links to SAP R/3 financial transactions as well as links to financial metrics for the company, stock market indices, e-mail, a calendar, and other information important for that person's job.

The advantage of having a personalized portal is its efficiency. A user only has to log on to one system to get all the information needed to perform a job. Without a portal, users often have to log on to multiple systems, such as an ERP system, industry exchanges, or suppliers' sites. Transferring information between systems is frequently difficult. With the Enterprise Portal, all information is available through the Web services provided by NetWeaver. Here, all the important links are presented in one screen, and transferring data is simplified by the ability to "drag-and-relate" data from one area to another.

Mobile Infrastructure is another module of NetWeaver. It allows users to access and work with data through mobile devices such as PDAs, cell phones, and pagers. Mobile Infrastructure provides remote access to data within the SAP system and other data contained within a company's information system. The benefits are obvious. A salesperson could use her PDA to see a customer's historical order information while in the middle of a sales visit at the customer's office. A partnership between SAP and VoiceObjects AG will add voice capability to NetWeaver, allowing users to enter data to the SAP CRM system using voice commands from their cell phones.[13]

Another facet of NetWeaver is Business Intelligence (BI). BI incorporates a data warehouse and data mining tools. BI can be delivered in a personalized manner with Enterprise Portal. It integrates information from various sources and processes both within and outside of the firm. BI works with any database management software and any operating system with NetWeaver.

Another NetWeaver feature, Master Data Management, provides data consistency within a company's SAP system. For example, in Fitter Snacker, the two sales groups, Direct and Wholesale, might have had different numbering systems for common customers. Master Data Management would ensure that the customer numbers are the same. The grocery industry could save $25 billion to $50 billion if suppliers could synchronize their data, such as product numbers, with retail outlets. NetWeaver allows this seamless Web interface to ensure proper data synchronization.[14]

NetWeaver's Exchange Infrastructure allows different applications to share data. By adhering to the standard of the Exchange Infrastructure, companies don't have to write code to enable different applications to transmit data. For example, using Exchange Infrastructure, a business can keep its current EDI system and seamlessly integrate that with its ERP system. SAP's Web Application Server, the development environment that is the foundation of NetWeaver, gives Exchange Infrastructure its customizability.[15]

Many non-users of NetWeaver are confused as to what it is. SAP admits the concept of tying applications together is not easy to explain in a few sentences in a marketing brochure. In response to this confusion, SAP has sponsored the *SAP NetWeaver for Dummies* publication and is also providing information sessions for their customers around the world.

Some customers that have installed NetWeaver do not expect results immediately. Kimberly-Clark Corporation hopes NetWeaver will allow it to check the status of inventory and orders using the Internet. At the start of its installation, Kimberly-Clark managers said this capability of NetWeaver is "definitely a vision, rather than reality".[16] Whirlpool has been happy with the benefits it has gained from NetWeaver in standardizing applications for its IT department. Rohm and Haas, a chemical firm, enjoys the portal aspect of NetWeaver.

NetWeaver at Work for Fitter Snacker

Now we will examine how NetWeaver could help Fitter Snacker. Assume that the two top salespeople, Amy Sanchez and Donald Brown, are busy selling NRG bars directly to customers and to distributors. Amy works from home. She logs on to the SAP system with her laptop computer using the SAP GUI. She doesn't know much about the SAP system, nor does she have to. She needs to know how to place customer orders and check on their status. When Amy goes on a sales call, she brings her notepad and calculator with her to jot down orders and quotes. When she returns home, she plugs those numbers into the SAP system and confirms her quotes. Amy would like to have some additional information on how salespeople in other regions are doing and what mix of bars they are promoting, but she doesn't know how to access any of that information. She also would like to see if there are new ways to market to her customers.

Donald Brown is also a salesperson, but he deals with distributors. He has been chosen to be a tester for the new NetWeaver SAP server. Every day, Donald comes into the office and logs on to his Enterprise Portal, which was tailored for his job. He sees figures from his top 10 customers, data on production and inventory of bars, the current stock quote for Fitter Snacker on the NASDAQ exchange, the current market price for oats, wheat germ, and honey, his e-mail, and the local weather report. Today Donald will make an important sales call for a regional grocery chain. He grabs his wireless PDA and some extra business cards, and heads out the door. During lunch with the purchasing agent for the grocery chain, Donald is able to check up-to-the-minute details on current sales orders and can confirm promises to ship additional bars next week, thanks to SAP's Mobile Infrastructure. Back at the office, Donald calls up the Business Intelligence module in NetWeaver. From there, he can run a few reports to find out which snack bars are currently selling better nationwide, grouped by region and time of year. He can also analyze snack bar sales using data mining to find sales patterns that can help him plan future sales calls.

Exercise 8.3

    a.   After reading about the features of SAP's NetWeaver and looking at how Amy Sanchez and Donald Brown perform their duties at Fitter Snacker, try to convince Fitter Snacker's CEO to implement NetWeaver. Write a memo to the CEO outlining your arguments.

    b.   The CEO is impressed with your work but has asked you for an ROI analysis (return on investment analysis). How do you begin doing that? What numbers do you need and who are the people you would have to interview to get those numbers?

ACCESSING ERP SYSTEMS OVER THE INTERNET

In addition to the standard GUI that is part of every ERP system, ERP vendors now offer access to their systems through a Web browser such as Microsoft Internet Explorer or Netscape Navigator. Users and systems administrators find it much more efficient to access their ERP systems through the browser; this method avoids the time-consuming installation of the GUI.

The University of Delaware's new PeopleSoft HR system was originally accessed through a PeopleSoft GUI. Since the University of Delaware processes most of its HR requests using forms on the Web, PeopleSoft agreed to create access to the system through a Web browser. Users can log on to the university's PeopleSoft system from home or from a remote office as long as they have a computer with an Internet connection and Web browser. They do not have to be in their own offices to perform their jobs. System administrators at the university also find that software upgrades are much simpler to manage. Only the software on the server needs to be upgraded—there is no special software installed on user PCs.

XML

Extensible Markup Language, shortened to **XML**, is the new programming language of the Internet. XML uses tags that define the data contained within them. Similar to assigning a data type to records in a database, XML tags apply specific meaning to the data within a Web page. XML-coded data can go directly from a Web page into a database without having to pass through middleware or, worse yet, be rekeyed into the system. In comparison, most Internet pages are written in Hypertext Markup Language, or HTML. HTML specifies only how your information will look (by assigning text styles, coloring, placement of graphics, and so on) when viewed through a browser.

XML users can create their own tags. Many companies are working together to create industry-specific tags to use with each other in conducting e-commerce. Fitter Snacker could keep its raw material records in XML format for easier transfer from suppliers' systems. One example of the XML document is shown in Figure 8-4. The customized tags in the document describe, or define, the data similar to how a database would. The ingredients honey and oats are defined as *descriptions* of an *item*.

FIGURE 8-4   Filter Snacker document in XML

Another example of the benefits of XML can be found in the insurance industry. Often in insurance, data must be passed between different documents. Sometimes the documents are from different companies. With XML, data are tagged and flow into those disparate documents without any additional processing. Every person involved in the chain of events—insurance brokers, insurance companies, underwriters, and insurance agents—can streamline their paper processing with XML.[17]

ERP systems are now ready to accept data in XML format. Using XML, companies can transfer data from their Web sites (data such as order information) directly into their ERP systems. This approach streamlines data entry, reduces errors, and reduces server loads.

XML is also very attractive to smaller companies. Small companies often transfer data over telephone lines or using fax machines. By using XML, electronic data transfer becomes much more affordable.[18] Without XML, a company might receive text-formatted data over the Internet. The data need to be converted from the text format to a database record—a potentially time-consuming and expensive process. Using XML tags to define the data would eliminate the conversion requirement.

Geon Incorporated, a polymer manufacturer, orders $250 million worth of products a year from OxyVinyls, its largest supplier.[19] Geon uses XML with its ERP system to integrate transactions in real time with OxyVinyls. Two servers (one at Geon and one at OxyVinyls) communicate through the Internet. A purchase order from Geon turns into an order entered at OxyVinyls without any person seeing the transaction. Inventory levels at Geon dropped by 6 percent the day the system went into effect, with further improvements expected.

ANOTHER LOOK

XML, ERP, and E-Commerce

Occidental Chemical Corporation, a subsidiary of Occidental Petroleum, uses SAP for its chemicals and resins business. When OxyChem linked its SAP system with a customer's SAP system, both companies realized tremendous customer service benefits. OxyChem has since embarked on a mission to link its SAP system with all of its customers. OxyChem has 5,000 customers of varying size in different industries. The company formed an internal team, which included customers, to study the project. The strategy they developed was to offer four different methods to integrate the systems.

The first method offers an XML-based ERP system-to-ERP system linkage between OxyChem and the customer. Once the two systems are linked, OxyChem can check the customer's ERP system to find out when it is becoming low on various raw materials. Purchase orders are automatically generated and orders flow efficiently.

The second method uses physical probes inside the customer's raw material containers. Using a Web interface and sales forecasts, the system automatically sends a purchase order in XML directly into OxyChem's SAP system.

In the third method, customers go to the ChemConnect chemical trading site (which you learned about earlier in this chapter) to order chemicals from OxyChem. These orders are created in XML, so they can flow directly into the SAP system at OxyChem.

The last, and most popular method, is an OxyChem-owned Web portal, that allows customers to place orders directly with OxyChem. The company originally thought this portal would be used only by its smaller customers, but customers of all sizes use it.

Of course, OxyChem still maintains a call center for those customers that do not want to order by computer.[20]

Questions:

1. Why did OxyChem give customers a choice in how to integrate their systems?
2. List the benefits of linking a manufacturer's ERP system with a customer's ERP system.
3. Draw a process map (see Chapter 7 for details on process mapping) to depict the flow of information from a manufacturing plant using a chemical from OxyChem. Show the process of requesting and then filling the order.

RADIO FREQUENCY IDENTIFICATION TECHNOLOGY

Radio frequency identification technology, known commonly as **RFID**, is becoming an efficient way of tracking items through a supply chain. An RFID device is a small package, or tag, that includes a microprocessor and an antenna that can be attached to products. The location of an item with an RFID tag can be determined using an RFID reader, which emits radio waves and receives signals back from the tag. The reader is also sometimes called an interrogator because it "interrogates" the tag. Since microprocessors continue to become more powerful and less expensive, according to Moore's Law discussed in Chapter 2, RFID technology has reached a point where it is inexpensive enough to be cost effective. Today most materials are tracked using bar codes and bar code readers. Bar code labels can degrade in bad weather, and an employee has to point a bar code reader at the bar code to read it. RFID technology does not need this line-of-site connection, and can withstand any environmental stresses.

Wal-Mart is in the process of implementing an RFID system to its supply chain. Suppliers will soon be shipping pallets of goods to Wal-Mart marked with RFIDs. Wal-Mart plans to read the RFID tags on cases and pallets when inventory enters a stockroom, when those cases of pallets go to the floor, and, ultimately, when empty cases are disposed. Much of the data collected during RFID reads will be passed on to Retail Link, Wal-Mart's Web-based software that lets its thousands of buyers and suppliers check inventory, sales, and more.[21]

Companies involved in manufacturing and selling RFID technology, like IBM, claim that their customers' projects experience a return on the RFID investment in less than 18 months.[22] Consumer goods companies are especially interested in this new technology. Companies such as Imperial Sugar and Black & Decker are implementing RFID technology.[23]

Procter & Gamble is using RFID technology to collect information about the sales of its products. The company uses its RFID-gathered data on fluctuations in sales of Pampers and Luvs diapers to help manage its supply chain processes. Although P&G sells disposable diapers to a population that is fairly constant, a slight change in the number of diapers sold at stores can be amplified by traditional supply chain planning processes, causing large fluctuations in demand throughout the supply chain, a phenomenon known as the **bullwhip effect**.[24] Assume that the number of diapers shipped to a large outlet chain remains fairly constant, but if there is a climate occurrence such as a hurricane, the normal supplies might be depleted. The bullwhip effect has been understood for a long time. The only way to alleviate this problem is to have better information for planning, which is what RFID technology promises to do. To figure out how many unsold diapers are in a particular store, someone has to go either count all the packages on the shelf and in the back room, or use a bar code reader and scan them. RFID tags eliminate the manual counting process. Determining the number of items in inventory in a building becomes automatic. With RFID, real-time inventory data are always available. RFID tags store detailed data, such as production lot number and production date, which provides even more benefit for items with an expiration date or that are subject to recall, like Fitter Snacker's line of NRG bars.

SAP's R/3 software is RFID ready. Along with partner company Infineon Technologies AG, SAP provides tools to easily link RFIDs to back-office systems. Through NetWeaver, SAP can integrate RFID data into both SAP and non-SAP applications.

Chapter Summary

1. E-commerce is transforming the way companies do business. Business-to-consumer e-commerce can streamline a company's ordering operations and record information about customers that can be used to customize sales and promotional activities, making the company more competitive.

2. Business-to-business e-commerce is changing the way companies buy and sell goods. New forms of procurement such as auctions, reverse auctions, and trading exchanges—all with dynamic pricing—are replacing the traditional intermediary.

3. ERP is an essential component for all forms of e-commerce. An integrated information system is required to provide speed and consistency in transaction processing and other back-office operations.

4. Application service providers (ASPs) are allowing companies to use ERP without a large initial investment, making ERP systems available to smaller companies. There are risks associated with using an ASP, however, and the decision to buy or lease must be weighed carefully.

5. Web services, a combination of software tools that lets various programs within an organization communicate with other applications, are gaining popularity.

6. SAP's Web services platform is NetWeaver, which includes tools for seamless connectivity of diverse applications through the World Wide Web. NetWeaver also includes modules such as Business Intelligence, Mobile Infrastructure, and Master Data Management.

7. Users of ERP systems often access those systems through a Web browser rather than the ERP systems' graphical user interface (GUI). Using a Web browser rather than a GUI program simplifies the software maintenance task.

8. XML, Extensible Markup Language, defines data on a Web page. ERP systems are using XML to integrate systems between suppliers and customers for easy data transfer.

9. RFID devices, or radio frequency identification devices, are used in tracking items in transit. RFIDs are particularly useful in supply chain processes for shipping and receiving cases and pallets of items. ERP vendors are developing the capability to incorporate RFID technology into ERP software.

Key Terms

Application service provider (ASP)	Exchange
Back-office processing	Extensible Markup Language (XML)
Bullwhip effect	Internet-based procurement
Business-to-business (B2B)	Portal
Business-to-consumer (B2C)	Radio frequency identification (RFID)
Commodity	Reverse auction
Denial of service (DoS)	Value-added network (VAN)
Electronic data interchange (EDI)	Web services

Exercises

1. Assume you have just graduated and you land a job at Fitter Snacker as its new purchasing agent. The person you replaced has just retired after 35 years in the job. You are eager to put to use some of the skills you learned at school, especially the work you did on B2B e-commerce. Write a memo to your new boss, the vice president of Supply Chain Management, on the virtues of e-commerce and why Fitter Snacker should now do Internet procurement.

2. Define application service provider. Do some research on the Web about current opinions of using ASPs. Try to find some examples of success and failure stories. Under what circumstances do you think using an ASP makes sense?

3. Go to the library and check out several publications about NetWeaver. Choose one aspect of NetWeaver and write a paper describing its functionality.

4. RFID technology is changing rapidly. Assume you have landed a summer job in the production department at Fitter Snacker. One of your jobs is to brief the entire team on the status of RFIDs. Create a PowerPoint presentation that brings the FS crew up to speed on this emerging technology.

End Notes

[1] "Europe E-Commerce to Grow Fourfold from 2003 to 2006," http://www.emarketer.com (July 2, 2003).

[2] Tobias Kollmann, "Measuring the Acceptance of Electronic Marketplaces: A Study Based on a Used-car Trading Site," *Journal of Computer-Mediated Communication* (2001).

[3] http://www.covisint.com.

[4] Julia Angwin, "Top Online Chemical Exchange Is an Unlikely Success Story," *The Wall Street Journal* (Jan. 8, 2004).

[5] Pimm Fox, "Private exchanges drive B2B success," *Computerworld* (May 7, 2001).

[6] "Let's Keep This Private," *Information Week* (Oct. 4, 2004).

[7] chemconnect.com white papers.

[8] "Learning Curve," http://www.darwinmag.com (June 14, 2001).

[9] Kathy Williams, "How Secure Is E-Commerce?" *Strategic Finance* (March 2000): 23.

[10] "Fast Track to Success," *Information Week* (June 21, 2004).

[11] http://www.rtr.com/Ready-to-Run_Software/ Web_Services-ROI_Notes1.pdf.

[12] Dan Woods and Jeffrey Word, *SAP NetWeaver for Dummies,* Wiley (2004).

[13] "SAP Adds Voice Capability," News Scan, *Information Week* (Sept. 9, 2004).

[14] SAP Solution Brief: SAP Master Data Management: http://www.sap.com/solutions/netweaver/masterdata/brochures/index.aspx.

[15] SAP Solution Brief: Exchange Infrastructure: http://www.sap.com/solutions/netweaver/exchangeinfrastructure/brochures/index.aspx.

[16] "SAP's NetWeaver Gets Tangled," *The Wall Street Journal* (online edition) (October 12, 2004).

17 Umar Kumesh Rai, White Paper: Realizing XML Benefits in Life Insurance, www.wipro.com/pdf_files/Wipro_XML_in_Insurance.pdf+benefits+of+xml+in+business&hl=en).

18 Marcus Herm, "XML - An opportunity for small and medium-sized enterprises," XML - The Site, http://www.softwareag.com/xml/library/herm.htm.

19 James E. Gaskin, "XML Comes of Age," InternetWeek.com (April 3, 2000).

20 "ERP+ERP=B2B," *Integrated Solutions* (July 2002).

21 Laurie Sullivan, "Wal-Mart's Way," *Information Week* (Sept. 27, 2004).

22 "IBM Takes RFID Services to Midsize Companies," *Information Week* (Sept. 9, 2004).

23 "RFID Tops To-do List in Consumer Goods," *Information Week* (Sept. 9, 2004).

24 "EZ Does It," *The New Yorker* (Sept. 8, 2003).

Accounting and Finance (A/F)   A functional area of business that is responsible for recording data about transactions, including sales, raw material purchases, payroll, and receipt of cash from customers.

Activity-based costing   An advanced form of inventory cost accounting in which overhead costs are assigned to products, based on the manufacturing activities that gave rise to the costs.

Advanced Business Application Programming (ABAP)   SAP R/3's internal programming language.

Application service provider (ASP)   A business that delivers software applications to companies over a network. Sometimes that network is the Internet.

Asset Management (AM) module   A module in SAP R/3 that helps a company to manage fixed-asset purchases (plant and machinery) and related depreciation.

Audit trail   A linked set of document numbers related to an order.

Back office processing   The processing of sales orders through a company's marketing, manufacturing, shipping, and accounting systems. A company selling on the Web must have an integrated back office to keep up with orders.

Balance sheet   A summary of a company's account balances that includes cash held, amounts owed to the company by customers, the cost of inventory on hand to be sold, long-term assets such as buildings, amounts owed to vendors, amounts owed to creditors, and amounts that the owners have invested in the company.

Best practices   The application of the best, most efficient ways in which business processes should be handled.

Bill of material (BOM)   The "recipe" listing the materials (including quantities) needed to make a product.

Bullwhip effect   Large fluctuations in demand throughout the supply chain caused by a slight change in the number of products sold.

Business functions   Business activities within a functional area of operation.

Business process   A collection of activities that takes one or more kinds of input and creates an output that is of value to a customer. Creating the output might involve activities from different functional areas.

Business process innovation (BPI)   The process of improving processes.

Business process reengineering   A quality improvement philosophy that recommends radical change to achieve radical improvements.

Business-to-business (B2B)   Communication and sales between manufacturers, wholesalers, retailers, and suppliers. This exchange can occur using both EDI and the Internet.

Business-to-consumer (B2C)   Communication and sales between businesses and the buying public. Popularly, but incorrectly, thought of as the most common form of e-commerce.

Cash-to-cash cycle time   The time from paying suppliers for raw materials to collecting cash from the customer (used in supply-chain management metrics).

Client-server architecture Data stored in a central computer (a server) are downloaded to a local PC (a client of the server) where data are processed. Historically, client-server architecture replaced many companies' mainframe-based architecture.

Commodity An item that is widely available at a standard level of quality.

Continuous improvement A quality improvement philosophy that prescribes systematic and repeated improvement efforts.

Controlling (CO) module A module in SAP R/3 that is used for internal management purposes. The software assigns a company's manufacturing costs to products and to cost centers, facilitating cost analysis.

Cost variance The difference between actual costs and standard costs.

Currency translation Converting financial-statement account balances expressed in one currency into balances expressed in another currency.

Customer master data Central database tables in SAP R/3 that store permanent data about each customer. Master data are used by many R/3 modules.

Customer relationship management (CRM) software A variety of different software tools that use data from a company's ERP system to enhance the company's relationships with its customers. CRM software allows these activities: segmenting customers, one-to-one marketing, sales-force automation (SFA), sales-campaign management, marketing encyclopedias, and call-center automation.

Data mining The statistical and logical analysis of large sets of transaction data, looking for patterns that can aid decision-making and improve customer sales and customer service. Data mining is often done with data in a data warehouse.

Data warehouse A database, separate from a company's operational database, that contains subsets of data from the company's ERP system. Users analyze and manipulate data in the warehouse. Thus, they do not interfere with the workings of the database that is used to record the company's transactions.

Database management system (DBMS) The technology that stores database records in an organized fashion and allows easy retrieval of the data.

Denial-of-service (DOS) attack A security attack on a Web site or Web server in which attackers block access to the Web-based service through a variety of means.

Deployment flowcharting (swimlane flowcharting) A type of flowchart that depicts team members across the top, with each process step aligned vertically under the employee or team working on it.

Development (DEV) system In a SAP system landscape, one of three separate SAP systems; DEV is used to develop configuration settings for the system using ABAP code.

Document flow The linked set of document numbers related to an order; an "audit trail."

Dynamic process modeling A method of evaluating process changes before they are implemented by putting into motion a basic process flowchart using computer simulation techniques to facilitate the evaluation of proposed process changes.

Electronic commerce (e-commerce) The buying and selling of goods and services over the Internet.

Electronic data interchange (EDI) A computer-to-computer transfer of standard business documents that allows companies to handle the purchasing process electronically, avoiding the cost and delays resulting from paper-based systems.

Enterprise Resource Planning (ERP) A collection of software systems that help to manage business processes such as marketing, production, purchasing, and accounting in an integrated way. ERP software does this by recording all transactions in a common database that is used by information systems throughout the company and by providing shared management-reporting tools.

Error log A record of discrepancies that occur during a payroll run.

Event process chain (EPC) A graphic model of a business process that uses only two symbols: events and functions.

Exchange A type of B2B electronic marketplace that focuses on a single industry.

Extensible markup language (XML) An Internet programming language that uses tags that define the data contained within them.

Financial Accounting (FI) module A module in SAP R/3 that records transactions in the general ledger accounts and generates financial statements for external reporting purposes.

Functional areas of operation A categorization of business activities, including marketing, sales, production, and accounting.

Gap analysis An assessment of disparities between an organization's current situation and its long-term goals.

General ledger A traditional record of accounting.

Hierarchical modeling The ability to flexibly describe a business process in greater or less detail depending on the task at hand.

Human capital management (HCM) Another term for Human Resources that describes the tasks associated with managing a company's workforce.

Human resources (HR) A functional area of business that manages recruiting, training, evaluation, and compensation of employees.

Human Resources (HR) module A module in SAP R/3 that facilitates employee recruiting, hiring, training, payroll, and benefits processing.

Income statement A financial statement that shows a company's profit or loss for a period of time.

Information system (IS) The computers, people, procedures, and software that store, organize, and deliver information.

Initial fill rate The percentage of the order that the supplier provided in the first shipment to the manufacturer or retailer (used in supply-chain management metrics).

Initial order lead time The time needed for the supplier to fill an order (used in supply-chain management metrics).

Integrated information system An information system that allows sharing common data throughout an organization. ERP systems are integrated systems because all operational data are located in a central database, where they can be accessed by users throughout an organization.

Intercompany transaction A transaction that occurs between a company and its subsidiary.

Internet-based procurement The use of Internet technologies for procurement activities.

Job A generic description of an employee's work responsibilities, or to the position that the person holds.

Legacy system An IS that is already in existence prior to installation of an ERP system.

Lead time The sum of the time it takes a manufacturer to get raw materials or subassemblies from a supplier—sum of the time required for a supplier to receive and process the order, take the material out of stock, package it, load it on a truck, and ship it to the manufacturer.

Lot sizing The process for determining purchase and production order quantities.

Managerial accounting Accounting that deals with determining the costs and profitability of the company's activities.

Marketing and sales (M/S) The functional area of business that is responsible for developing products, determining pricing, promoting products to customers, and taking customers' orders.

Master production schedule (MPS) The production plan for finished goods.

Materials Management (MM) A module in SAP R/3 that manages the acquisition of raw materials from suppliers (purchasing) and the subsequent handling and storage of raw materials, work in process, and finished goods.

Material Master Data Central database tables in SAP R/3 that store relatively permanent data about materials. These data are used by SD, MM, and other SAP R/3 modules.

Materials Requirements Planning (MRP) A production-scheduling methodology that determines the timing and quantity of production and purchase-order releases to meet a master production schedule. This process uses the bill of material, lot-size data, and material lead-times.

Metrics Measurements of performance; discussed in this book in relation to the effects of supply-chain management efforts.

Middleware Custom software that connects disparate systems from different vendors.

Modules Individual programs that can be purchased, installed, and run separately, but all extract data from the common database.

MRP Record The standard way of showing the Manufacturing Requirements Planning process on paper.

On-time performance A measure of how often a supplier meets agreed-upon delivery dates (used in supply-chain management metrics).

Open architecture Software that allows integration with third-party software. SAP R/3 is an example of open-architecture software. The term can also be applied to hardware products.

Organizational change management The supervision of human behavior aspects of organizational change.

Organizational structure The method used in SAP R/3 to define the relationships between organizational groups such as companies, plants, storage locations, sales divisions, and distribution channels.

Overhead A company's cost of operations, such as the costs for factory utilities, general factory labor, factory management, storage, insurance, and other manufacturing-related activities. Overhead is often called an indirect cost of production.

Payroll run The process of determining each employee's pay.

Plant Maintenance (PM) module A module in SAP R/3 that allows planning for preventive maintenance of plant machinery and managing maintenance resources, so equipment breakdowns are minimized.

Position A job that might exist in more than one department. For example, the administration assistant position might be required in the Wholesale Department as well as in Inside Sales.

Process boundary In process mapping, a definition of those activities to be included in the process, and what is considered part of the environment.

Process mapping A type of flowcharting that specifically represent pictorially the activities occurring within an *existing* business process.

Production (PROD) system In an SAP system landscape, one of three separate SAP systems; the actual system that the company uses to run its business processes.

Production Planning (PP) module A module in SAP R/3 that maintains production information; production is planned and scheduled, and actual production activities are recorded.

Project System (PS) module A module in SAP R/3 that allows planning for and control over new R&D, construction, and marketing projects. This module allows for costs to be collected against a project budget, and it can be used to manage the implementation of R/3 itself.

Qualifications Skills or abilities associated with a specific employee.

Quality assurance (QAS) system In a SAP system landscape, one of three separate SAP systems; the system where testing is done.

Quality Management (QM) module A module in SAP R/3 that helps to plan and record quality-control activities, such as product inspections and material certifications.

R/3 The 1992 version of the first integrated information system released by German software vendor SAP. This ERP system contains the following main modules, which can be implemented as a group or selectively: Sales and Distribution (SD), Materials Management (MM), Production Planning (PP), Quality Management (QM), Plant Maintenance (PM), Human Resources (HR), Financial Accounting (FI), Controlling (CO), Asset Management (AM), Project System (PS), Workflow (WF), and Industry Solutions (IS).

Radio frequency identification (RFID) A tracking technology that uses a small package, or tag, device that includes a microprocessor and antenna that can be attached to products. The location of an item with an RFID tag can be determined using an RFID reader, which emits radio waves and receives signals back from the tag.

Raw data Data on sales, manufacturing, and other operations that have not been analyzed.

Remuneration elements The part of an employee's pay that includes the base pay, bonuses, gratuities, overtime, sick pay and vacation allowances that the employee has earned during the pay period.

Repetitive manufacturing A manufacturing environment in which production lines are switched from one product to another similar product.

Requirements Skills or abilities associated with a position.

Return on investment (ROI) A ratio calculated by dividing the value of the project's benefits by the value of the project's cost.

Reverse auction An auction that features one buyer and many sellers.

Rough cut capacity planning The application of simple capacity estimating techniques to the production plans to see if they are feasible.

Sales and Distribution (SD) module A module in SAP R/3 that records sales orders and scheduled deliveries.

Sales forecast A company's estimate of future product demand, which is the amount of a product that customers will want to buy.

Sales forecasting The process of predicting the future demand for a company's products. Usually this is the first step in production planning.

Sales and operations planning The second step in the production planning process. At this step, data are evaluated to determine whether production facilities can produce enough goods to meet the consumer demand projected by the sales forecast.

Scalability The ability of information systems to be extended by adding servers to the network, rather than replacing the entire system. Scalability is a characteristic of client-server networks, but usually not of mainframe-based systems.

Scope creep The unplanned expansion of a project's goals and objectives, causing the project to go over time and over budget, as well as increasing the risk of an unsuccessful implementation.

Short list Up to three top candidates for a position, each of whom will be interviewed.

Software module See "module."

Standard cost The expected cost of manufacturing a product during a particular period. Standard costs for a product are established by (1) studying historical cost patterns in a company and (2) taking into account the effects of current manufacturing changes.

Statutory and voluntary deductions Paycheck withholdings that include taxes (federal, state, local, Social Security, and Medicare), company loans, and benefit contributions.

Stockout A manufacturing shortfall that occurs when raw materials or packaging run out.

Succession plan A company's plan for quickly installing a capable replacement if a key employee leaves the company.

Supply Chain Management (SCM) Sharing long-range production schedules between a manufacturer and its suppliers, so raw materials can be ordered and delivered in a timely manner, thus avoiding stockouts or excess inventory.

Tasks The assigned responsibilities related to a specific job.

Tolerance group Ranges that define limits on the dollar value of business transactions that an employee can process.

Value-added network (VAN) An EDI service or telecommunications provider. Companies acquire EDI service by subscribing to a VAN's EDI network.

Value analysis Analysis of each activity in a process for determining the value the activity adds to the product or service.

Web services A combination of software tools that lets various programs within an organization communicate with other applications.

Workflow (WF) A set of tools in SAP R/3 that can be used to automate any of the activities in SAP R/3. It can perform task-flow analysis and then prompt employees (by e-mail) if they need to take action.

Note: Page numbers in **boldface** denote pages where terms are defined.

D

E

EDI (electronic data interchange), **20**, 103–104, **192**

B2B e-commerce, 192–193

electronic commerce (e-commerce), **39**, 39–40, 191–211

business-to-business, **192**, 192–195

business-to-consumer, **192**

ERP, 195–196

electronic data interchange (EDI), **20**, 103–104, **192**

B2B e-commerce, 192–193

Ellison, Larry, 25

Enron, 124–127

Enterprise Portal, NetWeaver, 204

Enterprise Resource Planning (ERP), **1**

advantages, 17–18

applications, 33

continuing evolution, 38–40

degree of success using, 35–36

early attempts to share resources, 19–20

management's impetus to adopt, 21–22

manufacturing roots of, 20

maximizing value, 40

new directions, 24–28

significance, 32

EPCs (event process chains), **166**, 166–173

ERP. *See* Enterprise Resource Planning (ERP)

error logs, **153**

e-Synergy, 30

event process chains (EPCs), **166**, 166–173

Exact, 30

exchange(s), **193**, 193–194

Exchange Infrastructure, NetWeaver, 205

Extensible Markup Language (XML), **6**, 206–208

F

family leave management, SAP R/3, 156–157

Fastow, Andrew, 124, 125, 126

FedEx, 204

FI (Financial Accounting) module

general ledger entries, 112

SAP R/3, **27**

Financial Accounting, **109**, 109–110

Financial Accounting (FI) module

general ledger entries, 112

SAP R/3, **27**

financial statements, 109–110

financial transparency, 131–133

flowchart(s), **162**, 162–163

flowcharting

deployment, **166**

swimlane, **166**

Ford Motor Company, 193

FoxMeyer, 31, 35

fraudulent accounting records

Enron collapse, 124–127

Sarbanes-Oxley Act, 126–134

Freeman, Kenneth W., 155

Fuji Xerox Information Services (FXIS), 154

functional areas of operation, **2**, 2–3

information systems, 8–13

of very small businesses, 6–8

FXIS (Fuji Xerox Information Services), 154

G

gap analysis, **162**

Gates, Bill, 38

general ledger, **111**, 111–112

N

O

P

process improvement, 174–177

evaluating, 175–176

process mapping, **162**

extensions, 165–166

process model(s), **162**, 162–174

event process chain diagrams, 166–173

flowcharting, 162–163

process mapping extensions, 165–166

process modeling, dynamic, **176**

process-oriented management, 5–6

Procter & Gamble, 39, 155

PROD (Production) system, **185**

product profitability analysis, 117–122

companies with subsidiaries, 121–122

inaccurate inventory costing systems, 118–121

inconsistent record keeping, 117–118

production, 74–78

assemble-to-order approach, 74–75

make-to-order approach, 74

make-to-stock approach, 74

manufacturing process, 75–76

problems, 76–78

Production Planning (PP) module, SAP R/3, **27**

production planning process, 78–102

Demand Management, 79, 90–92

Detailed Scheduling, 79, 99–100

Materials Requirements Planning, 92–99

providing production data to accounting, 100–102

Sales and Operations Planning, 79, 81–90

Sales Forecasting, 79, 80–81

SAP R/3 approach, 79–80

Production (PROD) system, **185**

profitability analysis. *See* product profitability analysis

Project System (PS) module, SAP R/3, **27**

Prospecting phase, SAP CRM, 66

PS (Project System) module, SAP R/3, **27**

Purchasing, 80

purchasing problems, 77–78

Q

QAS (Quality Assurance) system, **185**

QM (Quality Management) module, SAP R/3, **27**

qualifications, **154**, 154–155

Quality Assurance (QAS) system, **185**

quality circles, **176**

Quality Management (QM) module, SAP R/3, **27**

Quest Diagnostics, 155–156

R

R/3 system. *See* SAP R/3

radio frequency identification (RFID), **9**, 209–210

raw data, **7**

record keeping, inconsistent, 117–118

recruiting process, 143–144

Reengineering the Corporation: A Manifesto for Business Revolution (Hammer), 21

Relational Software, Inc., 25

remuneration elements, **153**

repetitive manufacturing, **99**

reports, management. *See* management reporting

requirements, **154**

Retaining phase, SAP CRM, 66

return(s), 50